Inverse and Ill-Posed Problems Series

Coefficient Inverse Problems for Parabolic Type Equations and Their Application

Also available in the Inverse and Ill-Posed Problems Series:

An Introduction to Identification Problems via Functional Analysis
A. Lorenzi
Inverse Problems for Kinetic and other Evolution Equations
Yu.E. Anikonov
Inverse Problems of Wave Processes
A.S. Blagoveshchenskii
Uniqueness Problems for Degenerating Equations and Nonclassical Problems
S.P. Shishatskii, A. Asanov and E.R. Atamanov
Uniqueness Questions in Reconstruction of Multidimensional Tomography-Type Projection Data
V.P. Golubyatnikov
Monte Carlo Method for Solving Inverse Problems of Radiation Transfer
V.S. Antyufeev
Introduction to the Theory of Inverse Problems
A.L. Bukhgeim
Identification Problems of Wave Phenomena - Theory and Numerics
S.I. Kabanikhin and A. Lorenzi
Inverse Problems of Electromagnetic Geophysical Fields
P.S. Martyshko
Composite Type Equations and Inverse Problems
A.I. Kozhanov
Inverse Problems of Vibrational Spectroscopy
A.G. Yagola, I.V. Kochikov, G.M. Kuramshina and Yu.A. Pentin
Elements of the Theory of Inverse Problems
A.M. Denisov
Volterra Equations and Inverse Problems
A.L. Bughgeim
Small Parameter Method in Multidimensional Inverse Problems
A.S. Barashkov
Regularization, Uniqueness and Existence of Volterra Equations of the First Kind
A. Asanov
Methods for Solution of Nonlinear Operator Equations
V.P. Tanana
Inverse and Ill-Posed Sources Problems
Yu.E. Anikonov, B.A. Bubnov and G.N. Erokhin
Methods for Solving Operator Equations
V.P. Tanana
Nonclassical and Inverse Problems for Pseudoparabolic Equations
A. Asanov and E.R. Atamanov
Formulas in Inverse and Ill-Posed Problems
Yu.E. Anikonov
Inverse Logarithmic Potential Problem
V.G. Cherednichenko
Multidimensional Inverse and Ill-Posed Problems for Differential Equations
Yu.E. Anikonov
Ill-Posed Problems with A Priori Information
V.V. Vasin and A.L. Ageev
Integral Geometry of Tensor Fields
V.A. Sharafutdinov
Inverse Problems for Maxwell's Equations
V.G. Romanov and S.I. Kabanikhin

INVERSE AND ILL-POSED PROBLEMS SERIES

Coefficient Inverse Problems for Parabolic Type Equations and Their Application

P.G. Danilaev

///VSP///
UTRECHT • BOSTON • KÖLN • TOKYO
2001

VSP
P.O. Box 346
3700 AH Zeist
The Netherlands

Tel: +31 30 692 5790
Fax: +31 30 693 2081
vsppub@compuserve.com
www.vsppub.com

First published in 2001

ISBN 90-6764-349-3

Printed in The Netherlands by Ridderprint bv, Ridderkerk.

Contents

Preface

As a rule, many practical problems are studied in the situation when the input data are incomplete. For example, this is the case for a parabolic partial differential equation describing the non-stationary physical process of heat and mass transfer if it contains the unknown thermal conductivity coefficient. Such situations arising in physical problems motivated the appearance of the present work.

Coefficient inverse problems for parabolic equation are formulated as the problems of determining the unknown vector-function (Lavrent'ev, Romanov, and Shishatskii, 1980). Its components are represented by the function for which a given equation is written and the unknown coefficients of the elliptic differential operator entering this equation. Hereafter we shall assume that the unknown coefficients depend only on the space variables and do not depend on time.

Coefficient inverse problems were studied in Alifanov and Klibanov (1985), Vabishchevich and Denisenko (1990), and other works. Numerical methods of solution of coefficient inverse problems associated with their applications were developed in Abasov, Azimov, and Ibragimov (1991), Evdokimov (1995), Iskenderov (1971), Khairullin (1986, 1988), Tsirel'man (1984), Banks and Lamm (1985), Cahen (1963), Chavent (1970, 1971), Chavent, Dupuy, and Lemonnier (1975), Douglas and Jones (1962), Kravaris and Sienfeld (1982, 1985, 1986), etc.

Coefficient inverse problems are conditionally well-posed and require special methods of investigation. For example, these are the regularization method (Tikhonov and Arsenin, 1979; Tikhonov, Goncharskii, Stepanov, and Yagola, 1983; Morozov, 1987a,b), the quasi-solution method (Ivanov, Vasin, and Tanana, 1978), and the quasi-inversion method (as a version of the regularization method) (Lattès and Lions, 1967). Numerous works devoted to the method of solution of conditionally well-posed problems are well known today (see, for example, Bukhgeim, 2000; Lavrent'ev, Romanov, and

Shishatskii, 1980; Lavrent'ev, Reznitskaya, and Yakhno, 1982; Lavrent'ev, Vasil'ev, and Romanov, 1969; Lavrent'ev, 1981; Liskovets, 1981).

Although the regularization methods allow obtaining a stable solution according to the definition of well-posedness in the sense of A. N. Tihonov, this does not guarantee its uniqueness. The uniqueness of solution of conditionally well-posed problems was studied by M. M. Lavrent'ev, V. G. Romanov, and others. A wide variety of conditionally well-posed problems of mathematical physics that have practical applications are considered in Lavrent'ev, Romanov, and Shishatskii (1980), Romanov (1984), Romanov, Kabanikhin, and Pukhnacheva (1984), Romanov (1969), Belov and Lavrent'ev (1996). The specific feature of the present book is that it is based on the results obtained by M. V. Klibanov in the proofs of the corresponding uniqueness theorems (see Klibanov, 1984a,b, 1986; Klibanov and Danilaev, 1990). The main idea of these results consists consideration of the studied equation together with the overdetermined set of boundary conditions.

The quasi-inversion method developed by M. M. Lavrent'ev and J.-L. Lions was chosen as the method of solution. The developed algorithm allows reducing coefficient inverse problems to the problems on the continuation of a solution of a parabolic equation considered in Lattès and Lions, 1967. The quasi-inversion method was substantiated and developed further in Tamme (1972), Muzylev (1977), Popov and Samarskii, (1988), Vabishchevich (1991a,b,c), Samarskii and Vabishchevich (1997). M. Kh. Khairullin and M. N. Shamsiev used the quasi-inversion method to solve practical problems (see Shamsiev, 1997).

In this book we consider numerical solutions of the quasi-inversion problems, to which the solution of the original coefficient inverse problems are reduced. Numerical methods of solution of conditionally well-posed problems were developed by Bukhgeim (1986), Bakushinskii and Goncharskii (1989), Samarskii and Vabishchevich (1990), and others. The classic results obtained by A. A. Samarskii and his pupils were used to construct the algorithms of numerical solution of the quasi-inversion problems(see Samarskii, 1977; Samarskii and Nikolaev, 1978). Some studies presented in this book are associated with the numerical experiment as it was determined by A. A. Samarskii (Samarskii, 1979; Popov and Samarskii, 1988). The monographs of Alifanov (1979, 1988), Alifanov, Artyukhin, and Rumyantsev (1988), Beck, Blackwell, and Saint Clair (1989), Kozdoba and Krukovskii (1982), Kozdoba (1992), Kurpisz and Novak (1995) are devoted to applications of the methods of study of conditionally well-posed problems of mathematical physics. A wide range of practical applications of conditionally well-posed problems can be found in the papers of All-Union Seminars on Inverse problems conducted by Acad. A. N. Tihonov and Acad. V. P. Mishin, and in the proceedings of international conferences ("Identification of Dynamic Systems and Inverse Problems"), which are regularly held in the Aerospace Department of Moscow Institute of Aircraft Engineering under the direction of Prof. O. M. Alifanov.

Parabolic equations describe the processes of heat and mass transfer and are widely used in the mathematical modelling of physical processes (for example, Bolgarskii, Muhachev, and Shchyukin, 1975; Buzinov and Umrikhin, 1984; Bulygin, 1974; Charnyi, 1997; Golubev and Tumashev, 1972; Development of Research in Filtration Theory in the USSR (1917–1967), 1969; Kozdoba, 1975; Dmitriev, 1982; Sedov, 1967). In the present book, underground fluid dynamics is taken as a field of practical use of coefficient inverse problems. The significance of these problems for this application domain consists in the possibility to determine the physical fields of parameters that characterize the filtration properties of porous media (oil strata). This provides the possibility of predicting the conditions of oil-field development and the effects of the exploitation. Many authors proposed the algorithms of determining the fields of filtration parameters. In Bulygin (1958), Golubev, Danilaev, and Tumashev (1978), Khairullin (1986, 1988), Chavent (1970, 1971), Chavent, Dupuy, and Lemonnier (1975), Kravaris and Sienfeld (1982, 1985, 1986), this problem was solved using the methods of parametric identification with regularization. The role and specific character of the approach considered here as compared to the mentioned works are analyzed in Chapter 6. In the present book we generalized and developed the author's results (see Bulygin and Danilaev, 1971; Golubev, Danilaev, and Tumashev, 1978; Golubev 1992; Golubev and Danilaev, 1981, 1983, 1987, 1990, 1991a,b, 1992a,b, 1996a,b; Danilaev 1978, 1980, 1981, 1986, 1987a,b, 1988, 1989a,b, 1993, 1996, 1997, 1998a,b,; Danilaev and Golubev, 1994a,b; Danilaev, Gortyshov, and Kuz'min, 1988).

Chapter 1.

On the ill-posedness of coefficient inverse problems and the general approach to the study of them

We show the ill-posedness of coefficient inverse problems for parabolic equations following Lavrent'ev, Romanov, and Shishatskii (1980). Consider the equation

$$\frac{\partial u}{\partial t} - L_q u = f(x,t), \tag{1.1}$$

where L_q is a uniformly elliptic operator

$$L_q u = \sum_{i,j=1}^{2} a_{ij}(x) \frac{\partial u}{\partial x_i} \frac{\partial u}{\partial x_j} + \sum_{i=1}^{2} b_i(x) \frac{\partial u}{\partial x_i} + c(x)u,$$

$$a_{ij}(x) = a_{ji}(x), \qquad 0 < \mu_0 \le \sum_{i,j=1}^{2} a_{ij}(x)\alpha_i\alpha_j \le M_0 < \infty, \qquad \sum_{i=1}^{2} \alpha_i^2 = 1.$$

Here $q = (a_{11}, a_{12}, a_{22}, b_1, b_2, c)$ denotes the ordered set of the coefficients of the differential operator. Thus, defining the operator L_q is equivalent to specifying a vector q.

Let the elements q form some set Q associated with a set of operators L_q. From this set of operators we have to choose the one, which corresponds to the given information on the solution of equation (1.1). Let all components of the vector q be unknown. Then for their definition it is necessary to provide the data of dimension not less then that of the vector q. Consider, for example, the case where the number of solution of equation (1.1) is equal to the number of components of the vector q. Let these solutions be constructed as the solutions of initial boundary value problem corresponding to the different initial data. Introduce the vector function $U = (u_1, u_2, \dots, u_6)$ whose components consist of the solutions of equation (1.1) which conform to different initial data. For the function U we have the splitting system of Cauchy problems

$$U_t - L_q U = F(x,t), \qquad U|_{t=0} = U_0(x), \tag{1.2}$$

where the vector function U is given on some manifold $\mathcal{M}$:

$$U|_{\mathcal{M}} = \varphi(x,t). \tag{1.3}$$

It is required to find L_q, $q \in Q$, by φ.

Such statement of the problem is initial when we investigate the uniqueness and stability of solutions of coefficient inverse problems.

Let (q, U), $(\hat{u}, \hat{U})$ be two solutions of the problem (1.2) corresponding to the same functions F, U_0, but to the different data φ, $\hat{\varphi}$ in (1.3). Denote $\tilde{q} = q - \hat{q}$, $\tilde{U} = u - \hat{U}$, $\tilde{\varphi} = \varphi - \hat{\varphi}$. Then we obtain

$$\frac{\partial \tilde{U}}{\partial t} - L_q \tilde{U} = \tilde{q}(x) R_{\hat{q}}(x,t), \tag{1.4}$$

$$\tilde{U}|_{t=0} = 0, \tag{1.5}$$

$$\tilde{U}|_{\mathcal{M}} = \tilde{\varphi}(x,t), \tag{1.6}$$

where we use the notation

$$L_{\tilde{q}} \tilde{U} = \tilde{q}(x) R_{\hat{q}}(x,t).$$

Lemma (Lavrent'ev, Romanov, and Shishatskii, 1980). *The study of uniqueness and stability in the inverse problem* (1.2) *with respect to the data* (1.3) *is reduced to the study of analogous problems for determination of the function* $\tilde{q}(x)$ *from relations* (1.4)–(1.6), *where* L_q *is a given operator* ($q \in Q$); $\tilde{\varphi}(x,t)$ *and* $R_{\hat{q}}(x,t)$ *are given functions;* R *is a matrix.*

If we know the stability estimate for $\tilde{q}$ in terms of $\tilde{\varphi}$ which is uniform in Q, then this estimate is also a stability estimate for the solution of the problem (1.2), (1.3).

Thus, the problem of determining the differential operator L_q is reduced to the problem of determining the special right-hand side of the differential equation, as far as the study of uniqueness and stability is concerned. It is also an inverse problem, but it is linear in this case.

The questions of solution stability of the obtained problem depend on the function $R_{\hat{q}}(x,t)$. The requirement $\det R_{\hat{q}}(x,t) \not\equiv 0$ is intrinsic and minimal. The condition that $\det R_{\hat{q}}(x,t)$ is not equal to zero at all points of some smooth surface $t = t(x)$.

The linear problem (1.4)–(1.6) can be reduced to the Fredholm equation of the first kind using the fundamental solution $H_q(x,t,\xi,\tau)$ of equation (1.1):

$$\int_{\mathbb{R}^3} \tilde{q}(\xi) R_{q\hat{q}}(\xi,x,t)\,\mathrm{d}\xi = \tilde{\varphi}(x,t), \qquad (x,t) \in \mathcal{M}, \tag{1.7}$$

where

$$R_{q\hat{q}}(\xi,x,t) = \int_0^t R_{\hat{q}}(\xi,\tau) H_q(x,t,\xi,\tau)\,\mathrm{d}\tau.$$

Thus, the inverse problem (1.2), (1.3) is reduced to the integral Fredholm equation of the first kind (1.7).

The uniqueness theorems for solution of coefficient inverse problems for the parabolic equation (1.1) were proved by M. V. Klibanov. Now, we consider these theorems. Let $l > 0$ be a non-integer number (designations of Banach spaces conform to Ladyzenskaja, Solonnikov, and Ural'ceva (1967)), Ω be a bounded domain in $\mathbb{R}^n$, $\partial\Omega \in C^\infty$, $T = \text{const} > 0$, $Q_T = \Omega \times (0,T)$, $G_T = \Omega \times (-T,T)$, $G_{\sigma T} = \Omega \times (-\sigma, T)$, $\sigma = \text{const} \in (0,T)$, $S_T = \partial\Omega \times (0,T)$, $N_T = \partial\Omega \times (-T,T)$, L_q be a uniform elliptic operator with smooth coefficients in Q_T or G_T. The conditions of coordination of necessary order are assumed to hold in the further problems (Ladyzenskaja, Solonnikov, and Ural'ceva, 1967). Consider the linearized statements of inverse problems.

Problem 1.1. *Determine the vector function* $(u,q) \in H^{l,l/2}(\bar{Q}_T) \times H^{l-2}(\bar{\Omega})$, $l > 2$, *from the following conditions:*

$$\frac{\partial u}{\partial t} = Lu + q(x)F(x,t), \qquad (x,t) \in Q_T,$$

$$u|_{t=0} = 0, \qquad u|_{S_T} = \varphi(x,t), \quad \frac{\partial u}{\partial \nu}\Big|_{S_T} = \phi(x,t),$$

where L *is the uniformly elliptic operator with smooth coefficients in* Q_T *or* G_T, $L = \sum_{|\alpha| \le 2} a_\alpha D^\alpha$, $D^\alpha = D_1^{\alpha_1} \dots D_n^{\alpha_n}$, $D_j = \partial/\partial x_j$.

Problem 1.2. *Determine the functions u and q from the conditions of the previous problem with Q_T replaced by G_T.*

The uniqueness theorems follow from Klibanov (1984a, 1984b, 1986). For better understanding we shall present them as they were formulated in Klibanov and Danilaev (1990). The uniqueness theorems for the coefficient inverse problems follow from these theorems.

Theorem 1.1. *In Problem 1.2, suppose that $F(x,0) \neq 0$ in $\bar{\Omega}$, $l \in (4,5)$, $F \in H^{l,l/2}(\bar{G}_T)$, $a_\alpha \in H^{l-2,l/2-1}(\bar{G}_T)$. Then for all $T > 0$ there exists no more than one solution of Problem 1.2.*

The analogous theorem was proved for Problem 1.1.

Theorem 1.2. *Let $m = [n/2] + 3$, $l \in (m, m+1)$, $a_\alpha, b \in H^l(\bar{\Omega})$, $F(x,0) \neq 0$ in $\bar{\Omega}$, $F(x,t) \in R(A)$, $A^{-1}(F) \in H^l(\bar{Q}_{T_1})$ $\forall T_1 > 0$,*

$$L = \sum_{k=1}^{n} \frac{\partial}{\partial x_k}\left(\frac{1}{c(x)}\frac{\partial}{\partial x_k}\right) + \sum_{|\alpha|\leq 1} a_\alpha(x) D^\alpha = L_0 + \sum_{|\alpha|\leq 1} a_\alpha(x) D^\alpha,$$

$$\mu_1 \leq \frac{1}{c} \leq \mu_2, \qquad \inf_{\Omega}\left(2c(x) + |x|\frac{\partial c}{\partial |x|}\right) > 0.$$

Then, for all $T > 0$, $f \equiv \varphi \equiv \phi \equiv 0$ implies $u \equiv b \equiv 0$.

In Problem 1.1 we assume that $F \neq 0$ in $\bar{Q}_T$ and the assumptions of Theorem 1.2 are hold. For $\lambda =$ const denote

$$\tilde{u} = \mathrm{e}^{-\lambda t}\,\frac{u}{F}$$

$$v = \frac{\partial \tilde{u}}{\partial t} + \lambda \tilde{u} \Rightarrow \int_0^t \mathrm{e}^{-\lambda(t-T)} v(x,\tau)\,\mathrm{d}\tau.$$

We obtain

$$v_t - \tilde{L}v - \sum_{j=1}^{n} \int_0^t K_j(x,t,\tau)\mathrm{e}^{-\lambda(t-\tau)} D_j v(x,\tau)\,\mathrm{d}\tau = 0,$$

$$v|_{S_T} = \hat{\varphi}, \qquad \left.\frac{\partial v}{\partial \nu}\right|_{S_T} = \hat{\phi}, \tag{1.8}$$

where $\hat{\varphi}$, $\hat{\phi}$, K_j are known functions; K_j does not depend on λ; $\tilde{L}$ is an elliptic operator with the main part L_0. For $\varepsilon_1, \varepsilon_2 > 0$, $\varepsilon = (\varepsilon_1, \varepsilon_2)$ consider the boundary value problem (the quasi-inversion problem)

$$Pv_\varepsilon - \varepsilon_2 \frac{\partial^2 v_\varepsilon}{\partial t^2} + \frac{1}{\varepsilon_1} P^* P v_\varepsilon = 0,$$

$$-v_\varepsilon(x,0)\varepsilon_2\,\frac{\partial v_\varepsilon}{\partial t}(x,0)+\frac{1}{\varepsilon_1}(Pv_\varepsilon)(x,0)=0,$$

$$\varepsilon_2\,\frac{\partial v_\varepsilon}{\partial t}(x,T)+\frac{1}{\varepsilon_1}(Pv_\varepsilon)(x,T)=0,$$

$$v_\varepsilon|_{S_T}=\hat{\varphi},\qquad \frac{\partial v_\varepsilon}{\partial \nu}\Big|_{S_T}=\hat{\phi}, \tag{1.9}$$

where P is an operator in the left-hand part of (1.8). Multiplying both parts of (1.9) by the function $\eta\in\mathring{W}_2^{2,1}$, $\partial\eta/\partial\nu|_{S_T}=0$ and integrating with respect to Q_T, we obtain

$$\begin{aligned}\int_{Q_T}\Big[\sum_{k=1}^{n}\frac{1}{c(x)}\frac{\partial v_\varepsilon}{\partial x_k}\frac{\partial \eta}{\partial x_k}-v_\varepsilon\frac{\partial\eta}{\partial t}+\varepsilon_2\frac{\partial v_\varepsilon}{\partial t}\frac{\partial\eta}{\partial t}\\ +\eta(\tilde{L}-L_0)(v)+\frac{1}{\varepsilon_1}(Pv_\varepsilon)(P\eta)\Big]\,\mathrm{d}x\,\mathrm{d}t\\ +\int_\Omega v_\varepsilon(x,T)\eta(x,T)\,\mathrm{d}x=0.\end{aligned} \tag{1.10}$$

Definition. The function $v_\varepsilon\in W_2^{2,1}$ which satisfies (1.10) and the conditions for S_T which enter the statement of this problem is called a generalized solution of the quasi-inversion problem.

Theorem 1.3. *Suppose that there exists a function $\Phi\in W_2^{2,1}$ such that*

$$\Phi|_{S_T}=\hat{\varphi},\qquad \frac{\partial\Phi}{\partial\nu}\Big|_{S_T}=\hat{\phi}.$$

Then there exits a number $\lambda>0$ depending on the properties of the operator L and the function F such that there exits a unique generalized solution of the quasi-inversion problem, $\|v_\varepsilon\|_{2,1}\le C(\varepsilon)\|\Phi\|_{2,1}$, where the constant $C(\varepsilon)$ does not depend on v_ε and Φ. If the assumptions of Theorem 1.2 hold for $m_1=m+2$ and the solution of Problem 1.2 exists, then

$$\lim_{\varepsilon_1\to 0}\Big[\sum_{k=1}^{n}\Big\|\frac{\partial v_\varepsilon}{\partial x_k}-\frac{\partial v}{\partial x_k}\Big\|_{L_2(Q_T)}+\|v_\varepsilon(x,t)-v(x,t)\|_{L_2(\Omega)}\Big]=0,$$

for $\varepsilon_2=\text{const}$ and for all $t\in(0,T)$ and there exists a rule according to which $\sigma=\sigma(\varepsilon_1)\to 0$ so that $\|v_\varepsilon(x,\sigma)-b(x)\|_{L_2(\Omega)}\to 0$. If the function Φ is given with error δ in the norm $W_2^{2,2}$ and $\varepsilon_1=\delta^\alpha$, $\alpha=\text{const}\in(0,2)$, then the limit equality will also hold as $\delta\to 0$.

Problem 1.2 should be solved in the same way as Problem 1.1, but $t = 0$ in the boundary condition in the statement of the quasi-inverse problem should be replaced by $t = -\sigma$, and Q_T should be replaced by $G_{\sigma T}$.

Theorem 1.4. *In Problem 1.2 suppose that $F \geq 1$ in G_T, the assumptions of Theorem 1.1 hold, and there exists a function Φ described in Theorem 1.3. Suppose $\|u\|_{H^{l,l/2}(\bar{G}_T)} + \|F\|_{H^{l,l/2}(\bar{G}_T)} \leq M$, $\varepsilon_2 = \text{const} \in (0,1)$. Then there exist numbers λ, σ, γ depending on M, ε_2 and the properties of L (we can estimate them explicitly: the number σ is quite small) such that if the solution of Problem 1.2 exists, then the following estimate is valid:*

$$\|v_\varepsilon - v\|_{W_2^{2,1}(G_\sigma)} + \|v_\varepsilon(x,0) - b(x)\|_{L_2(\Omega)} \leq C\varepsilon_1^\gamma,$$

where the constant C does not depend on ε_1, v, and b.

Thus, we have to define the vector function

$$(\tilde{q}, U) \in H^{l,l/2}(\bar{Q}_T) \times H^{l-2}(\bar{\Omega}), \quad l > 2,$$

from the conditions

$$\frac{\partial U}{\partial t} = LU + \tilde{q}(x)R(x,t), \qquad (x,t) \in Q_T,$$

$$U|_{t=0} = 0,$$

$$U|_{S_T} = \varphi(x,t), \qquad \left.\frac{\partial U}{\partial \nu}\right|_{S_T} = \psi(x,t)$$

to solve the problem (1.2) by the data (1.3).

This problem has overdetermined boundary conditions and is conditionally well-posed. In this paper, we use the quasi-inversion method as one of the methods of regularization to solve the coefficient inverse problems (Lattes and Lions, 1970).

Chapter 2.

Determining the coefficient of the lowest term of equation

2.1. SETTING OF THE PROBLEM. DETERMINATION OF THE COEFFICIENT

It is required to find the coefficient $q(x)$ of the lowest term of one-dimensional parabolic equation

$$u_t - u_{xx} = q(x)u, \tag{2.1}$$

where the unknown coefficient $q(x)$ does not depend on the variable t.

We formulate this problem as a coefficient inverse problem following Klibanov (1984a, 1984b, 1986), Klibanov and Danilaev (1990). Determine the vector function $\{q(x), u(x,t)\}$ from the conditions

$$\begin{gathered} u_t - u_{xx} = q(x)u, \qquad 0 < x < 1, \quad t > 0, \\ u(0,t) = f_0(t), \qquad u_x(0,t) = f_1(t), \qquad t > 0; \\ u(1,t) = g_0(t), \qquad u_x(1,t) = g_1(t), \qquad t > 0; \\ u(x,0) = u_0(x), \qquad 0 \leq x \leq 1. \end{gathered} \tag{2.2}$$

Transform equation (2.1) into the form, that does not contain the unknown coefficient. For this purpose we solve this equation with respect to the coefficient q and differentiate the obtained equality with respect to t observing that $q_t = 0$:

$$q = \frac{u_t - u_{xx}}{u}, \tag{2.3}$$

$$u_{tt}u - u_{xxt}u - u_t^2 + u_{xx}u_t = 0. \tag{2.4}$$

In equation (2.4) we pass from the sought function $u(x,t)$ to the new unknown function $\nu = u_t/u$. The following the relations are true:

$$\frac{u_{tt}}{u} = v^2 + v_t, \qquad \frac{u_{xxt}}{u} = \frac{u_{xx}}{u}v + 2\frac{u_x}{u}v_x + v_{xx},$$

$$\frac{u_x}{u} = \int_0^t v_x \, dt + \frac{u_{0x}}{u_0}.$$

As a result, we have the solution of the nonlinear integro-differential equation with the corresponding set of additional conditions

$$v_t - v_{xx} = 2v_x\left[\int_0^t v_x(x,t)\, dt + \frac{u_{0x}}{u_0}\right], \qquad 0 < x < 1, \quad t > 0;$$

$$\begin{aligned} &v(0,t) = \eta_0(t), \qquad v_x(0,t) = \eta_1(t), \qquad t > 0, \\ &v(1,t) = \mu_0(t), \qquad v_x(1,t) = \mu_1(t), \qquad t > 0, \end{aligned} \tag{2.5}$$

the initial condition $v(x,0)$ is not set.

The problem (2.5) is conditionally well-posed. It contains the overdetermined set of boundary conditions and does not contain the initial condition. We use the quasi-reversibility method to solve this problem (Lattes and Lions, 1970; Klibanov and Danilaev, 1990; Danilaev, 1989, 1997; Samarskii and Vabishchevich, 1997; Vabishchevich, 1991,a,b).

Deduce the equation of quasi-reversibility method which corresponds to the obtained integro-differential equation. As applied to the considered problem, it has the following form

$$\frac{1}{\varepsilon_1} P^* P\tilde{v} + P\tilde{v} - \varepsilon_2 \tilde{v}_{tt} = 0.$$

In this connection, we have to construct an operator which is conjugate to the nonlinear integro-differential operator. Assume

$$P\tilde{v} = \tilde{v}_t^{(n)} - \tilde{v}_{xx}^{(n)} - 2\tilde{v}_x^{(n-1)}\left(\int_0^t \tilde{v}^{(n-1)}(x,t)\, dt + \frac{u_{0x}}{u_0}\right),$$

where n is the number of iteration. Then we obtain

$$P^*\tilde{v} = -\tilde{v}_t^{(n)} - \tilde{v}_{xx}^{(n)}.$$

As a result, we arrive at the following problem of the quasi-inversion method

$$-\frac{1}{\varepsilon_1}\tilde{v}_{tt} + \frac{1}{\varepsilon_1}\tilde{v}_{xxxx} + \tilde{v}_t - \tilde{v}_{xx} - \varepsilon_2\tilde{v}_{tt} = f(x,t) - \frac{1}{\varepsilon_1}f_t(x,t) - \frac{1}{\varepsilon_1}f_{xx}(x,t), \tag{2.6}$$

$$\begin{aligned} &\tilde{v}(0,t) = \eta_0(t), \qquad \tilde{v}_x(0,t) = \eta_1(t), \qquad t > 0, \\ &\tilde{v}(1,t) = \mu_0(t), \qquad \tilde{v}_x(1,t) = \mu_1(t), \qquad t > 0, \\ &\varepsilon_2\tilde{v}(x,t_1) + \frac{1}{\varepsilon_1}(\tilde{v}_t(x,t) - \tilde{v}_{xx}(x,t_1) - f(x,t_1)) = 0, \end{aligned}$$

where

$$f(x,t) = 2\tilde{v}_x^{(n-1)}\left(\int_0^t \tilde{v}_x^{(n-1)}(x,t)\,\mathrm{d}t + \frac{u_{0x}}{u_0}\right), \qquad 0 < x < 1, \quad t > 0.$$

The initial condition $\tilde{v}(x,0)$ is arbitrary, which is admissible in the quasi-inversion method. Besides, the usual conditions of concordance are observed and the following notations is used:

$$\eta_0(t) = \frac{f_{0t}(t)}{f_0(t)}, \qquad \eta_1(t) = \frac{f_{1t}(t)f_0(t) - f_1(t)f_{0t}(t)}{f_0^2(t)}$$

$$\mu_0(t) = \frac{g_{0t}(t)}{g_0(t)}, \qquad \mu_1(t) = \frac{g_{1t}(t)g_0(t) - g_1(t)g_{0t}(t)}{g_0^2(t)}.$$

After solving the quasi-inversion problem, i. e., equation (2.6) with the corresponding additional conditions, we return to find the unknown coefficient $q(x)$. Rewrite equation (2.1) in the form (2.3) and calculate the expressions

$$\frac{u_x}{u} = \int_0^t \tilde{v}_x(x,t)\,\mathrm{d}t + \frac{u_{0x}}{u_0},$$

$$\begin{aligned} \frac{u_{xx}}{u} &= \left(\frac{u_x}{u}\right)^2 + \int_0^t \tilde{v}_{xx}(x,t)\,\mathrm{d}t + \frac{u_{0xx}}{u_0} - \left(\frac{u_{0x}}{u_0}\right)^2 \\ &= \left(\int_0^t \tilde{v}_x(x,t)\,\mathrm{d}t\right)^2 + 2\frac{u_{0x}}{u_0}\int_0^t \tilde{v}_x(x,t)\,\mathrm{d}t \\ &\quad + \int_0^t \tilde{v}_{xx}(x,t)\,\mathrm{d}t + \frac{u_{0xx}}{u_0}, \end{aligned}$$

using the above change of variables $v = (\partial u/\partial t)/u$. Substitute these relations into (2.3). We obtain

$$q(x) = \tilde{v}(x,t) - \left(\int_0^t \tilde{v}_x(x,t)\,\mathrm{d}t\right)^2 - 2\frac{u_{0x}}{u_0}\int_0^t \tilde{v}_x(x,t)\,\mathrm{d}t$$
$$- \int_0^t \tilde{v}_{xx}(x,t)\,\mathrm{d}t - \frac{u_{0xx}}{u_0}. \tag{2.7}$$

We shall consider the right-hand side of the formula (2.7) for some fixed $t = t_k$. If the time interval t_k is small, that is, formula (2.7) is simplified as follows:

$$q(x) \approx \tilde{v}(x,t_k) - \frac{u_{0xx}}{u_0}. \tag{2.8}$$

2.2. THE DIFFERENCE QUASI-INVERSION PROBLEM

We consider the network function $\tilde{v}_i^n = \tilde{v}(ih, n\tau)$ on the mesh with the steps $h = 1/J$ and $\tau = t_1/N$. Here J and N are numbers of steps with respect to the variables x and t, correspondingly. Equation (2.6) in the quasi-inversion problem is nonlinear. Construct the iterative process to solve this equation. For the sought function $(\tilde{v}^{(k)})_i^n$ we introduce the additional upper index k to designate the iteration number. We also use the notation $(u_0)_i = u_0(ih)$.

Take advantage of the finite differences of second order accuracy. Introduce the column vector of the unknowns,

$$\bar{V}_n^{(k)} = \begin{pmatrix} (\tilde{v}^{(k)})_1^n \\ (\tilde{v}^{(k)})_2^n \\ \cdot \\ (\tilde{v}^{(k)})_{J-1}^n \end{pmatrix}.$$

We shall write the difference equation of the quasi-inversion problem in the matrix form

$$\mathbf{A}\bar{V}_{n-1}^{(k)} - \mathbf{C}\bar{V}_n^{(k)} + \mathbf{B}\bar{V}_{n+1}^{(k)} = -\bar{D}_n^{(k-1)}, \qquad n = 1,\dots,N-1, \tag{2.9}$$

where

$$\mathbf{A} = -\Big(\frac{1}{\tau^2}\Big(\varepsilon_2 + \frac{1}{\varepsilon_1}\Big) + \frac{1}{2\tau}\Big)\mathbf{E}_{J-1},$$

$$\mathbf{B} = -\Big(\frac{1}{\tau^2}\Big(\varepsilon_2 + \frac{1}{\varepsilon_1}\Big) - \frac{1}{2\tau}\Big)\mathbf{E}_{J-1},$$

E is the unit matrix of order $(J-1)\times(J-1)$,

$$\mathbf{C}=\begin{pmatrix} -(P+G) & -N & -M & 0 & . & 0 & 0 & 0 & 0 \\ -Q & -P & -N & -M & . & 0 & 0 & 0 & 0 \\ -G & -Q & -P & -N & . & 0 & 0 & 0 & 0 \\ 0 & -G & -Q & -P & . & 0 & 0 & 0 & 0 \\ . & . & . & . & . & . & . & . & . \\ 0 & 0 & 0 & 0 & . & -G & -Q & -P & -N \\ 0 & 0 & 0 & 0 & . & 0 & -G & -Q & -(P+M) \end{pmatrix}_{(J-1)\times(J-1)},$$

$$P=\frac{2}{\tau^2}\Big(\varepsilon_2+\frac{1}{\varepsilon_1}\Big)+\frac{6}{\varepsilon_1 h^4}+\frac{2}{h^2},$$

$$N=Q=-\Big(\frac{4}{\varepsilon_1 h^4}+\frac{1}{h^2}\Big),\qquad M=G=\frac{1}{\varepsilon_1 h^4},$$

$$\bar{D}_n^{(k-1)}=\begin{pmatrix} -(F^{(k-1)})_1^n-\Big(\frac{4}{\varepsilon_1 h^4}+\frac{1}{h^2}\Big)\eta_0^n+\frac{2h\eta_1^n}{\varepsilon_1 h^4} \\ -(F^{(k-1)})_2^n+\frac{\eta_0^n}{\varepsilon_1 h^4} \\ -(F^{(k-1)})_3^n \\ . \\ -(F^{(k-1)})_{J-3}^n \\ -(F^{(k-1)})_{J-2}^n+\frac{\mu_0^n}{\varepsilon_1 h^4} \\ -(F^{(k-1)})_{J-1}^n-\Big(\frac{4}{\varepsilon_1 h^4}+\frac{1}{h^2}\Big)\mu_0^n+\frac{2h\mu_1^n}{\varepsilon_1 h^4} \end{pmatrix}_{(J-1)\times 1},$$

$$\begin{aligned}(F^{(k-1)})_i^n=&\frac{1}{2h^2}((\tilde{v}^{(k-1)})_{i+1}^n-(\tilde{v}^{(k-1)})_{i-1}^n)\\ &\times\Big(\tau\sum_{n=0}^{n}((\tilde{v}^{(k-1)})_{i+1}^n-(\tilde{v}^{(k-1)})_{i-1}^n)+\frac{u_{0\,i+1}-u_{0\,i-1}}{u_{0\,i}}\Big)\\ &-\frac{1}{4\varepsilon_1 h^2}\Big(((\tilde{v}^{(k-1)})_{i+1}^{n+1}-(\tilde{v}^{(k-1)})_{i-1}^{n+1})\sum_{n=0}^{n}((\tilde{v}^{(k-1)})_{i+1}^{n+1}-(\tilde{v}^{(k-1)})_{i-1}^{n+1})\\ &-((\tilde{v}^{(k-1)})_{i+1}^{n-1}-(\tilde{v}^{(k-1)})_{i-1}^{n-1})\sum_{n=0}^{n}((\tilde{v}^{(k-1)})_{i+1}^{n-1}-(\tilde{v}^{(k-1)})_{i-1}^{n-1})\Big)\end{aligned}$$

$$
\begin{aligned}
&-\frac{1}{4\varepsilon_1\tau h^2}\big((\tilde v^{(k-1)})_{i+1}^{n+1}-(\tilde v^{(k-1)})_{i-1}^{n+1}-(\tilde v^{(k-1)})_{i+1}^{n-1}+(\tilde v^{(k-1)})_{i-1}^{i+1}\big)\\
&\qquad\times\frac{u_{0\,i+1}-u_{0\,i-1}}{u_{0\,i}}\\
&-\frac{\tau}{2\varepsilon_1 h^4}\Big(((\tilde v^{(k-1)})_{i+2}^{n}-(\tilde v^{(k-1)})_{i}^{n})\sum_{n=0}^{n}((\tilde v^{(k-1)})_{i+2}^{n}-(\tilde v^{(k-1)})_{i}^{n})\\
&\qquad-2((\tilde v^{(k-1)})_{i+1}^{n}-(\tilde v^{(k-1)})_{i-1}^{n})\sum_{n=0}^{n}((\tilde v^{(k-1)})_{i+1}^{n}-(\tilde v^{(k-1)})_{i-1}^{n})\\
&\qquad+((\tilde v^{(k-1)})_{i}^{n}-(\tilde v^{(k-1)})_{i-2}^{n})\sum_{n=0}^{n}((\tilde v^{(k-1)})_{i}^{n}-(\tilde v^{(k-1)})_{i-2}^{n})\Big)\\
&-\frac{1}{2\varepsilon_1 h^4}\Big(((\tilde v^{(k-1)})_{i+2}^{n}-(\tilde v^{(k-1)})_{i}^{n})\frac{u_{0\,i+2}-u_{0\,i}}{u_{0\,i+1}}\\
&\qquad-2((\tilde v^{(k-1)})_{i+1}^{n}-(\tilde v^{(k-1)})_{i-1}^{n})\frac{u_{0\,i+1}-u_{0\,i-1}}{u_{0\,i}}\\
&\qquad+((\tilde v^{(k-1)})_{i}^{n}-(\tilde v^{(k-1)})_{i-2}^{n})\frac{u_{0\,i}-u_{0\,i-2}}{u_{0\,i-1}}\Big).
\end{aligned}
$$

The first boundary condition of equation (2.9) has the form

$$-\mathbf{C}_0\bar V_0^{(k)}+\mathbf{B}_0\bar V_1^{(k)}=-\bar D_0. \tag{2.10}$$

In relation (2.10) we shall assume that $\mathbf{C}_0=\mathbf{E}_{J-1}$ and $\mathbf{B}_0$ is the zero matrix in view of the initial condition $\tilde v(x,0)=v^*(x)$ where $v^*(x)$ is an arbitrary function. Then we obtain the equality

$$\bar V_0^{(k)}=\bar D_0$$

or

$$\begin{pmatrix}(\tilde v^{(k)})_1^0\\ \cdot\\ (\tilde v^{(k)})_{J-1}^0\end{pmatrix}=\begin{pmatrix}v_1^*\\ \cdot\\ v_{J-1}^*\end{pmatrix},$$

where $v_i^*=v^*(ih)$.

The second boundary condition has the form

$$\mathbf{A}_N\bar V_{N-1}^{(k)}-\mathbf{C}_N\bar V_N^{(k)}=-\bar D_N^{(k-1)}, \tag{2.11}$$

where

$$\mathbf{A}_N=-\tau^{-1}(\varepsilon_2+\varepsilon_1^{-1})\mathbf{E}_{J-1},$$

$$b=\tau^{-1}(\varepsilon_2+\varepsilon_1^{-1})+2/\varepsilon_1h^2,\qquad a=c=-1/\varepsilon_1h^2,$$

$$\mathbf{C}_N = \begin{pmatrix} -b & -a & 0 & . & 0 & 0 & 0 \\ -c & -b & -a & . & 0 & 0 & 0 \\ 0 & -c & -b & . & 0 & 0 & 0 \\ . & . & . & . & . & . & . \\ 0 & 0 & 0 & . & -c & -b & -a \\ 0 & 0 & 0 & . & 0 & -c & -b \end{pmatrix}_{(J-1)\times(J-1)},$$

$$\bar{D}_N^{(k-1)} = \begin{pmatrix} -(\varphi^{(k-1)})_1^N - \dfrac{\eta_0^N}{\varepsilon_1 h^2} \\ -(\varphi^{(k-1)})_2^N \\ . \\ -(\varphi^{(k-1)})_{J-2}^N \\ -(\varphi^{(k-1)})_{J-1}^N - \dfrac{\mu_0^N}{\varepsilon_1 h^2} \end{pmatrix}_{(J-1)\times 1},$$

$$(\varphi^{(k-1)})_i^n = \frac{1}{2h^2}((\tilde{v}^{(k-1)})_{i+1}^N - (\tilde{v}^{(k-1)})_{i-1}^N)$$
$$\times \left(\tau \sum_{n=0}^{N} ((\tilde{v}^{(k-1)})_{i+1}^n - (\tilde{v}^{(k-1)})_{i-1}^n) + \frac{u_{0\,i+1} - u_{0\,i-1}}{u_{0\,i}}\right).$$

Write the solution procedure for the system of vector equations, which is usual for the sweep method (see Samarskii, A.A. and Nikolaev, E.S. (1978)). Taking into account the form of the coefficients of vector equations, we have

$$\alpha_{n+1} = (\mathbf{C} - \mathbf{A}\alpha_n)^{-1}\mathbf{B}, \qquad n = 1, \dots, N-1, \quad \alpha_1 = 0,$$
$$\bar{\beta}_{n+1} = (\mathbf{C} - \mathbf{A}\alpha_n)^{-1}(\mathbf{A}\bar{\beta}_n + \bar{D}_n^{(k-1)}), \qquad n = 1, \dots, N, \quad \bar{\beta}_1 = \bar{D}_0,$$
$$\bar{V}_N^{(k-1)} = (\mathbf{C}_N - \mathbf{A}_N\alpha_N)^{-1}(\mathbf{A}_N\bar{\beta}_N^{(k-1)} + \bar{F}_N^{(k-1)}),$$
$$\bar{V}_n^{(k)} = \alpha_{n+1}\bar{V}_{n+1}^{(k-1)} + \bar{\beta}_{n+1}^{(k-1)}, \qquad n = N-1, N-2, \dots, 1, 0.$$

The initial approximation to start the iterations is

$$\bar{V}_n^{(0)} = \bar{V}_N^{(0)} = \bar{D}_0 = \begin{pmatrix} v_1^* \\ . \\ v_{J-1}^* \end{pmatrix}.$$

The iteration process stops under the usual condition

$$|\bar{V}_n^{(k)} - \bar{V}_n^{(k-1)}| \leq \varepsilon,$$

where ε is a small prescribed constant.

2.3. A TEST EXAMPLE

The following test example was constructed to estimate the effectiveness of the proposed algorithm:

$$u_t - u_{xx} = xu, \qquad (x,t) \in (0,1) \times (0,T),$$

$$u(0,t) = \exp\left(\frac{1}{3}t^3\right), \qquad u_x(0,t) = t\exp\left(\frac{1}{3}t^3\right),$$

$$u(1,t) = \exp\left(t + \frac{1}{3}t^3\right), \qquad u_x(1,t) = t\exp\left(t + \frac{1}{3}t^3\right), \qquad t > 0,$$

$$u(x,0) = 1, \qquad x \in [0,1].$$

The example has the exact analytical solution

$$u(x,t) = \exp\left(xt + \frac{1}{3}t^3\right).$$

The corresponding function $v(x,t)$ is a solution to the following problem

$$v_t - v_{xx} = 2v_x \int_0^t v_x \,\mathrm{d}t, \qquad (x,t) \in (0,1) \times (0,T],$$

$$v(0,t) = t^2, \qquad v_x(0,t) = 1, \qquad t \in (0,T],$$

$$v(1,t) = 1 + t^2, \qquad v_x(1,t) = 1, \qquad t \in (0,T],$$

$$v(x,0) = x, \qquad x \in [0,1] \tag{2.12}$$

The exact solution of this problem is $v(x,t) = x + t^2$.

In calculations, the initial condition of the subsidiary problem (2.12) is replaced by an arbitrary initial condition (here $v(x,0) = 0$) according to the general procedure of the method and the general statement of the problem.

We chose the mesh domain in the following way. The number of steps with respect to the variable x did not change: $J = 10$, the step size $h = 0.1$. The number of steps with respect to t was changing during the computation. The step size τ and the time of determination of the computation T, were changing as well. Different values of the regularization parameters ε_1 and ε_2 were chosen in the process of computation in accordance with previous recommendations (Danilaev, 1989, 1997, 1987, 1993). The computation results are presented in the diagrams below. In Figures 2.1–2.11, the calculated values of $\tilde{v}(x,t)$ (dashed curve) are compared with the theoretical values (continuous line). The values of the parameters ε_1, ε_2, τ, N, and T

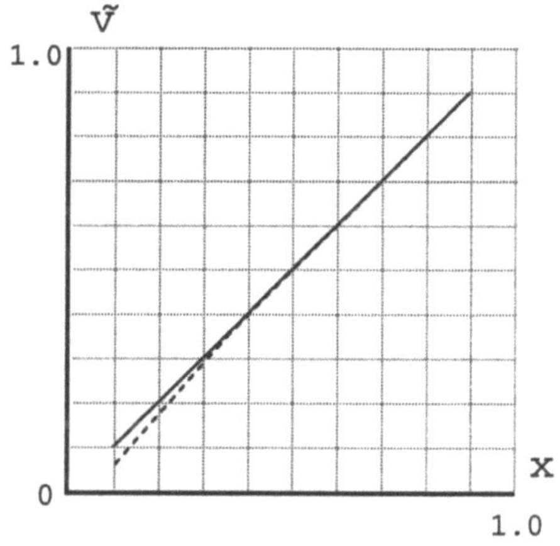

Figure 2.1: Parameters:
$\varepsilon_1 = 0.8$, $\varepsilon_2 = 10^{-3}$,
$\tau = 10^{-4}$, $T = 10^{-2}$

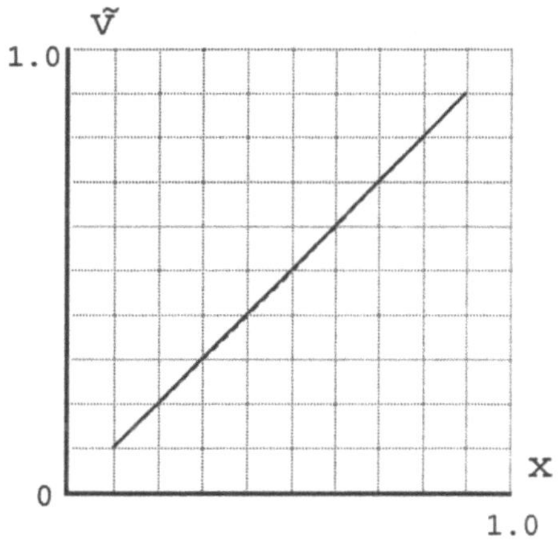

Figure 2.2: Parameters:
$\varepsilon_1 = 0.8$, $\varepsilon_2 = 10^{-3}$,
$\tau = 10^{-4}$, $T = 10^{-3}$.

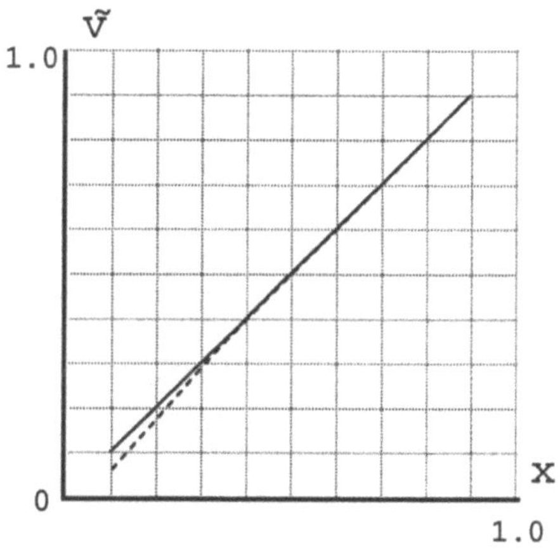

Figure 2.3: Parameters:
$\varepsilon_1 = 8 \cdot 10^{-3}$, $\varepsilon_2 = 10^{-3}$,
$\tau = 10^{-3}$, $T = 10^{-2}$

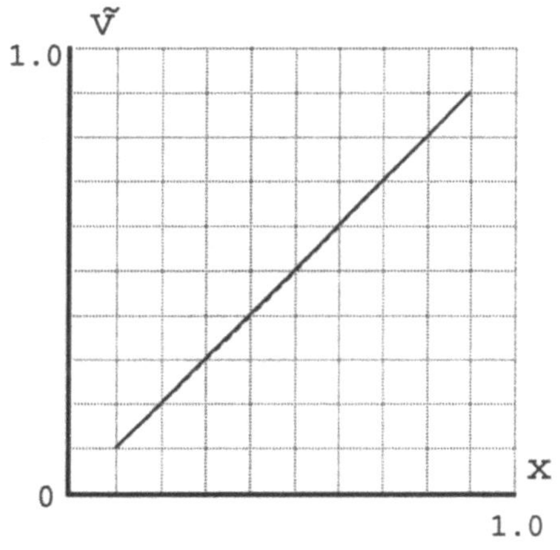

Figure 2.4: Parameters:
$\varepsilon_1 = 81$, $\varepsilon_2 = 10^{-3}$,
$\tau = 10^{-5}$, $T = 10^{-4}$

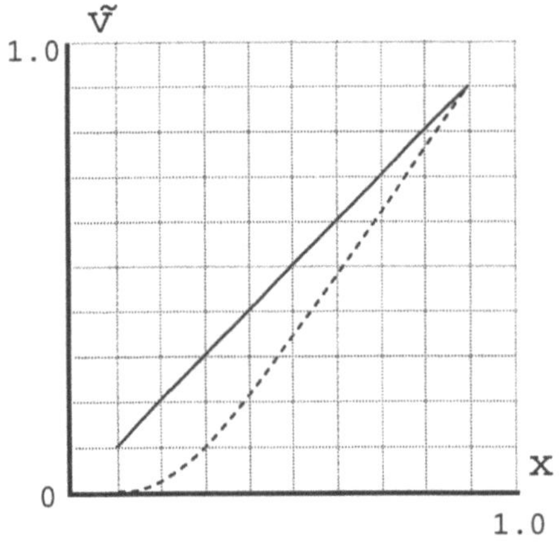

Figure 2.5: Parameters:
$\varepsilon_1 = 8 \cdot 10^{-3}$, $\varepsilon_2 = 10^{-3}$,
$\tau = 10^{-3}$, $T = 10^{-1}$

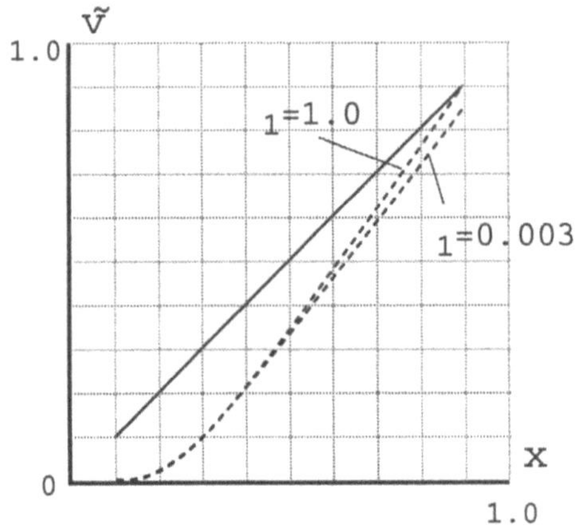

Figure 2.6: Parameters:
$\varepsilon_2 = 10^{-3}$,
$\tau = 10^{-2}$, $T = 10^{-1}$

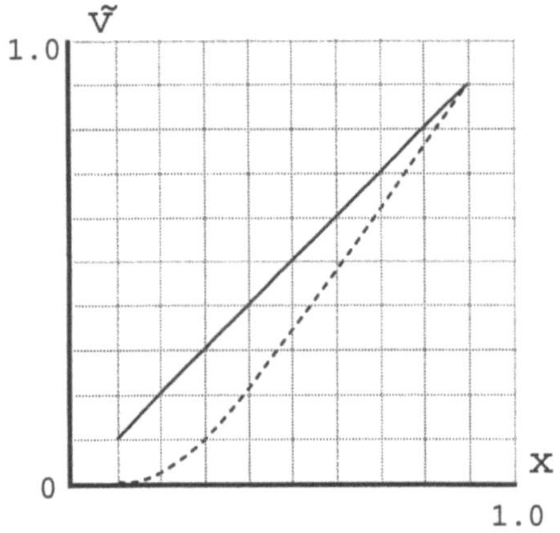

Figure 2.7: Parameters:
$\varepsilon_1 = 1.0$, $\varepsilon_2 = 10^{-3}$,
$\tau = 10^{-2}$, $T = 10^{-1}$

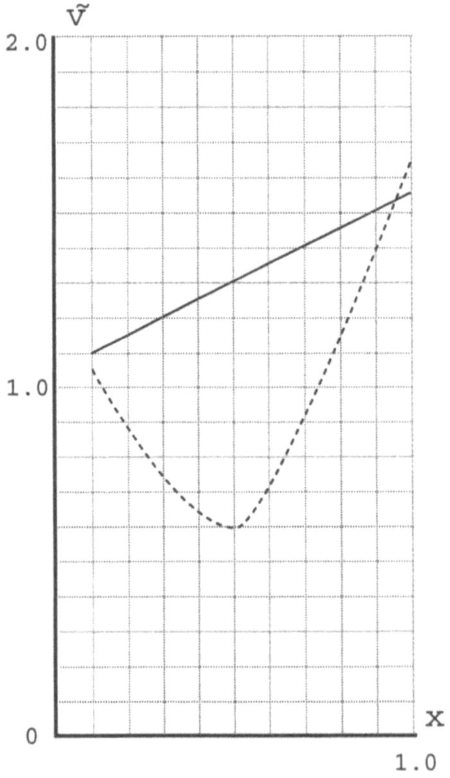

Figure 2.8: Parameters:
$\varepsilon_1 = 1.0$, $\varepsilon_2 = 10^{-3}$,
$\tau = 10^{-2}$, $T = 10^{-1}$

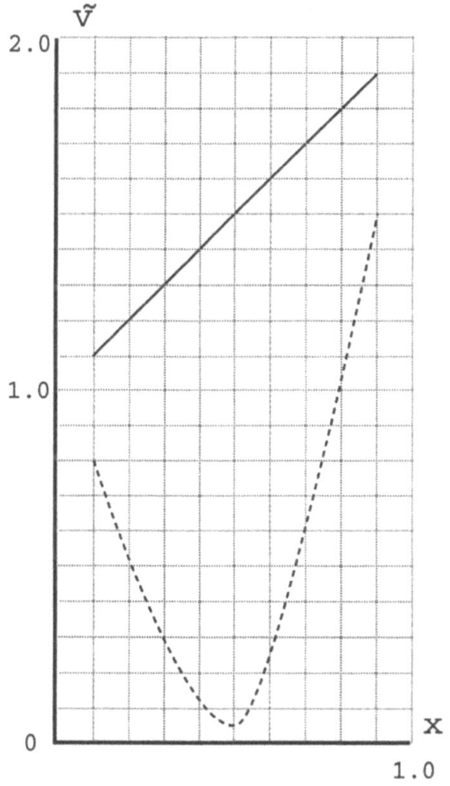

Figure 2.9: Parameters:
$\varepsilon_1 = 3 \cdot 10^{-4}$, $\varepsilon_2 = 10^{-3}$,
$\tau = 10^{-2}$, $T = 1.0$

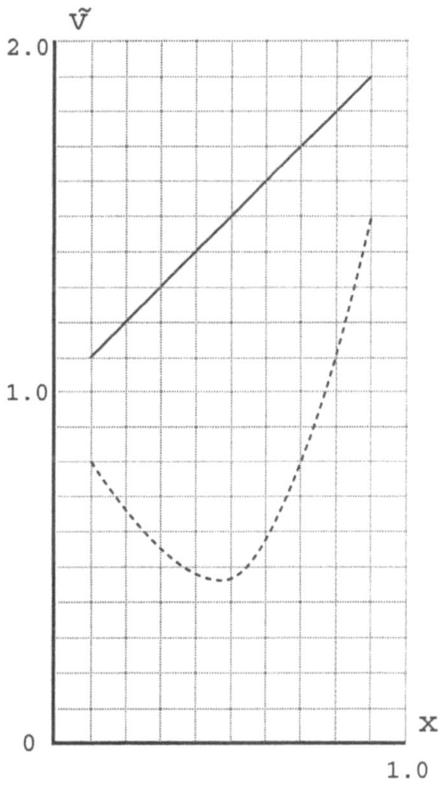

Figure 2.10: Parameters:
$\varepsilon_1 = \cdot 10^{-4}$, $\varepsilon_2 = 10^{-3}$,
$\tau = 10^{-1}$, $T = 1.0$

corresponding to the calculated values of $\tilde{v}(x,t)$ are given for each diagram. The initial approximation of $\tilde{v}(x,t)$ was chosen to be zero in all cases.

The computation results are typical for the general way of applying the quasi-reversibility method for numerical solution of parabolic equations. The characteristic features are the same as those described before in Danilaev (1997, 1989, 1993), Danilaev and Golubev (1994). The results are worse for large values of T. In this case the calculations should be terminated when $t_1 < T$ and the result should be used for further solution of the quasi-reversibility problem. The computation results are scarcely affected by the regularization parameter ε_2. They improve with the increase of the other regularization parameter, ε_1 (up to a certain limit, above which the computation results do not change anywhere).

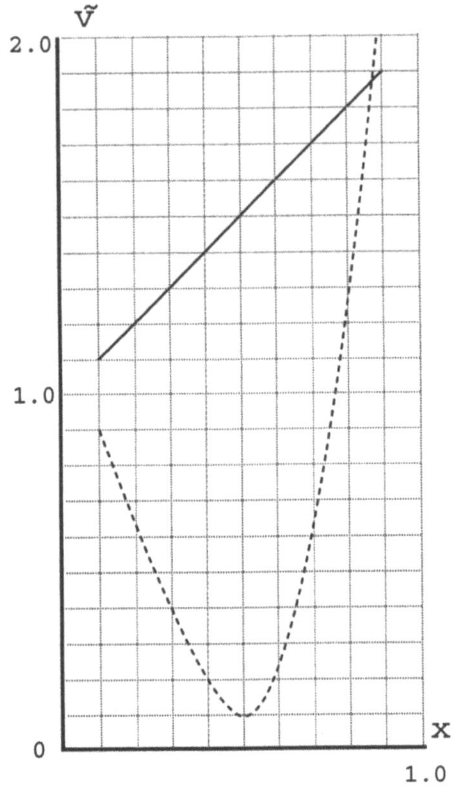

Figure 2.11: Parameters:
$\varepsilon_1 = 1.0$, $\varepsilon_2 = 10^{-3}$,
$\tau = 10^{-2}$, $T = 1.0$

Another computational experiment was carried out to study the influence of perturbations of boundary conditions on the solutions stability. The relative error in setting the boundary conditions in the C^2 norm did not exceed 5 % and scarcely affected the results.

Further solution of the problem consisted in determining the unknown functions $q(x)$ and $u(x,t)$. According to formula (2.8), for the present test example we have $q(x) \approx \tilde{v}(x,t_k)$. Thus, Figures 2.1–2.11 can be viewed as the comparison of the numerical values of the unknown coefficient $q(x)$ (dashed curve)and the values given by the test problem conditions (continuous line) because the exact solution, $v(x,t)$, and $q(x)$ coincide for small t. The value of t_k for calculation of the sought coefficient by formula (2.8) was chosen on the basis of the usual conditions of the quasi-reversibility method for the problem on the extension of a solution of a parabolic equation. The interval between this value and the termination point of the computation was 2–3 time steps.

As an example, we give the termination condition for the iteration process of the computation illustrated by Figure 2.1:

$$|(\tilde{v}^{(k+1)})_j^n - (\tilde{v}^{(k)})_j^n| \leq 0.01 \sum_{j=1}^{J} |(\tilde{v}^{(k)})_j^n| \, J^{-1}.$$

It was satisfied after five iterations. The problem required 1 minute of computation time for the computer ES-1033. Figures 2.2 and 2.4 demonstrate that the numerical solution and the exact solution almost coincide. The character of deviation of the numerical solution from the exact solution is typical for the quasi-reversibility method. The accuracy of reconstruction of the coefficient $q(x)$ dramatically decrease for $T > 0.1$. This can be explained by the nonlinearity of the problem.

Chapter 3.

Determining of the coefficient for the leading terms of equation

3.1. STATEMENT OF THE PROBLEM

We study the problem of determining the coefficient $k(x)$ which enters the divergent principle part of the parabolic equation

$$(ku_x)_x = u_t \tag{3.1}$$

and does not depend on time. The set of additional conditions necessary for obtaining the unique solution of this coefficient inverse problem is determined by the corresponding uniqueness theorem (see Klibanov and Danilaev, 1990). Dimension of the information which is used according to the requirements of the theorem satisfies the conditions specified in Chapter 1.

Consider the coefficient inverse problem for equation (3.1) written in the form with expanded divergent principal part:

$$ku_{xx} + k_x u_x = u_t. \tag{3.2}$$

The coefficient inverse problem for equation (3.1) or, equivalently, for equation (3.2) as follows can be formulated.

Determine the vector function $\{k(x), u(x,t)\}$ solving the following problem:

$$ku_{xx} + k_x u_x = u_t, \qquad x_0 < x < x_1, \quad -\varepsilon < t \leq t_1,$$

$$
\begin{gathered}
u(x_0,t) = g_0(t), \qquad u_x(x_0,t) = h_0(t), \\
u(x_1,t) = g_1(t), \qquad u_x(x_1,t) = h_1(t), \qquad -\varepsilon < t \le t_1, \\
u(x,0) = \varphi(x), \qquad x_0 \le x \le x_1.
\end{gathered} \tag{3.3}
$$

The ordinary conditions of concordance hold:

$$
q_0(0) = \varphi(x_0), \qquad g_1(0) = \varphi(x_1), \qquad h_0(0) = \varphi_x(x_0), \qquad h_1(0) = \varphi_x(x_1).
$$

Eliminate the unknown coefficient $k(x)$ from equation (3.2) as proposed in (Lavrent'ev, Romanov, and Shishatskii, 1980). The transformation is based on the following idea.

Suppose that an equation has the form $f(x,t) = 0$ and there exists a function $F(x,t)$ which is primitive with respect to t for the function $f(x,t)$. Then it is evident that the following operations are valid:

$$
\int_0^t f(x,t)\,\mathrm{d}t = 0, \quad F(x,t) - F(x,0) = 0, \quad (F(x,t) - F(x,0))_t = f(x,t).
$$

Further we use these equalities in order to avoid high order of the partial differential equation to be obtained as a result of the equation transformation.

Integrate equation (3.2) with respect to time and write the result to solve for the coefficient $k(x)$:

$$
k = \frac{(u-\varphi) - k_x \displaystyle\int_0^t u_x\,\mathrm{d}t}{\displaystyle\int_0^t u_{xx}\,\mathrm{d}t}. \tag{3.4}
$$

We use the fact that the coefficient k does not depend on t, i.e. $k_t = 0$. Differentiate equality (3.4) with respect to t. Write the result to solve for the derivative k_x:

$$
k_x = \frac{(u-\varphi)u_{xx} - u_t \displaystyle\int_0^t u_{xx}\,\mathrm{d}t}{u_{xx}\int_0^t u_x\,\mathrm{d}t - u_x \int_0^t u_{xx}\,\mathrm{d}t}. \tag{3.5}
$$

Repeat the operation of differentiation with respect to t. As follows from the condition $k_{xt} = 0$, after differentiation of equality (3.5) we obtain the equation

$$
\begin{aligned}
&(u-\varphi)(u_{xx}u_{xt} - u_{xxt}u_x) \\
&\quad = \int_0^t u_{xx}\,\mathrm{d}t(u_{xt}u_t - u_x u_{tt}) + \int_0^t u_x\,\mathrm{d}t(u_{xx}u_{tt} - u_{xxt}u_t).
\end{aligned} \tag{3.6}
$$

Note that such operations are proposed in (Lavrent'ev, Romanov, and Shishatskii, 1980).

Replace the sought function $u(x,t)$ in equation (3.6) by the new unknown function $v(x,t) = u_t/u$. After the change we obtain the nonlinear integro-differential equation with the corresponding set of additional conditions:

$$Av_{xx} + Bv_x = Cv_t, \qquad x_0 < x < x_1, \quad -\varepsilon < t \le t_1, \tag{3.7}$$

$$v(x_0,t) = \mu_0(t), \qquad v_x(x_0,t) = \eta_0(t), \qquad -\varepsilon < t \le t_1, \tag{3.8}$$

$$v(x_1,t) = \mu_1(t), \qquad v_x(x_1,t) = \eta_1(t), \qquad -\varepsilon < t \le t_1, \tag{3.9}$$

with the initial condition for $v(x,0)$ being arbitrary. Here we use the following notation:

$$\mu_0(t) = \frac{g_{0t}}{g_0}, \qquad \mu_1(t) = \frac{g_{1t}}{g_1}, \qquad \eta_0(t) = \frac{h_{0t}g_0 - h_0 g_{0t}}{g_0^2},$$

$$\eta_1(t) = \frac{h_{1t}g_1 - h_1 g_{1t}}{g_1^2}, \quad A = EJ_0 + vJ_1, \quad B = E(2J_0^2 - J_2) + v(2J_0J_1 - J_3),$$

$$C = J_1J_2 - J_0J_3, \qquad E = 1 - \exp\Big(\int_0^t v\,\mathrm{dt}\Big),$$

$$J_0 = \int_0^t v_x\,\mathrm{dt} + (\ln\varphi)_x, \qquad J_1 = \int_0^t (1-E)J_0\,\mathrm{dt},$$

$$J_4 = \int_0^t v_{xx}\,\mathrm{dt} + \frac{\varphi_{xx}}{\varphi}, \qquad J_2 = J_0^2 - \Big(\frac{\varphi_x}{\varphi}\Big)^2 + J_4, \qquad J_3 = \int_0^t (1-E)J_2\,\mathrm{dt}.$$

The problem (3.7)–(3.9) differs from the classical problem. It is conditionally well-posed and requires specific methods of solution. We use the quasi-inversion method for solving this problem.

3.2. THE QUASI-INVERSION PROBLEM AND AN ESTIMATE OF STABILITY OF ITS SOLUTION

Write the equation of the quasi-inversion method for the considered problem (3.7)–(3.9)

$$\frac{1}{\varepsilon_1}P^*P\tilde{v} + P\tilde{v} - \varepsilon_2\tilde{v}_{tt} = 0, \tag{3.10}$$

where

$$P\tilde{v} = -A\tilde{v}_{xx} - B\tilde{v}_x + C\tilde{v}_t,$$

P^* is the conjugate of P. Equation (3.10) is considered with the corresponding set of initial-boundary conditions (Lattes and Lions, 1970). The quasi-inversion method consists in constructing of a new problem whose solution $v(x,t)$ is close to the solution of the original problem in a certain metric, which is proved below.

We solve the problem of constructing the operator conjugate to the nonlinear integro-differential operator as follows:

$$P\tilde{v}^{(n)} = -A^{(n-1)}(\tilde{v}^{(n)})_{xx} - B^{(n-1)}(\tilde{v}^{(n)})_x + C^{(n-1)}(\tilde{v}^{(n)})_t,$$

where n is the iteration number. Then we define the conjugate operator as usual:

$$P^*\tilde{v}^{(n)} = -(C\tilde{v})_t + (B\tilde{v})_x - (A\tilde{v})_{xx}.$$

Under these assumptions, equation (3.10) takes the form

$$\begin{aligned}
&\frac{1}{\varepsilon_1}A^2\tilde{v}_{xxxx} + \frac{1}{\varepsilon_1}4AA_x\tilde{v}_{xxx}\\
&+\Big\{\frac{1}{\varepsilon_1}[2A_x^2 + 2AA_{xx} - B^2 + (AB)_x + (AC)_t] - A\Big\}\tilde{v}_{xx}\\
&+\Big\{\frac{1}{\varepsilon_1}[-2BB_x + (BC)_t + (AB)_{xx}] - B\Big\}\tilde{v}_x\\
&+\Big\{\frac{1}{\varepsilon_1}[-2CC_t - (AC)_{xx} + (BC)_x] + C\Big\}\tilde{v}_t\\
&-\Big(\frac{1}{\varepsilon_1}C^2 + \varepsilon_2\Big)\tilde{v}_{tt} + \frac{1}{\varepsilon_1}[-2(AC)_x + 2BC]\tilde{v}_{tx} = 0.
\end{aligned} \tag{3.11}$$

We study the questions of stability of a solution of the quasi-inversion problem following Samarskii and Vabishchevich (1997). Consider the problem

$$\frac{dU}{dt} - AU = 0, \qquad U(0) = U_0.$$

Determine the approximate solution of this problem from the solution of the quasi-inversion problem

$$\frac{dU_\alpha}{dt} - AU_\alpha + \alpha A^*AU_\alpha = 0$$

and the initial conditions $U_\alpha(0) = U_0$.

The following estimate of stability is obtained in Samarskii and Vabishchevich (1997):

$$\|U_\alpha(t)\|^2 \le \exp\Big(\frac{1}{2\alpha}t\Big)\, \|U_\alpha(0)\|^2.$$

In this work, the results concerning the stability with respect to the initial data are formulated for the unstable problem with operators that are not self-conjugate ($\mathrm{A} \neq \mathrm{A}^*$). The problem

$$\frac{\mathrm{d}U}{\mathrm{d}t} - \mathrm{A}U = 0, \qquad U(0) = U_0$$

is considered in a Hilbert space H. Here A is a constant independent of t linear operator with the domain of definition $D(\mathrm{A})$ which is dense in H. It is supposed that A is positive definite in H and, in general, unbounded.

Let $\mathrm{A} \neq \mathrm{A}^*$ be a constant operator. For determining $u_\alpha(t)$ use the equation

$$\frac{\mathrm{d}u_\alpha}{\mathrm{d}t} - \mathrm{A}u_\alpha + \alpha\, \mathrm{A}^*\mathrm{A}u_\alpha = 0.$$

Apply the following theorem to estimate the stability of a solution of the perturbed problem.

Theorem 3.0 (Samarskii and Vabishchevich, 1997). *Let $D = G^*G > 0$ and let the constant operator G be commuting with* A *and* A^*. *Then for the solution of the problem*

$$u_\alpha(0) = u_0, \qquad \frac{\mathrm{d}u_\alpha}{\mathrm{d}t} - \mathrm{A}u_\alpha + \alpha\, \mathrm{A}^*\mathrm{A}u_\alpha = 0$$

the following estimate is valid:

$$\|u_\alpha(t)\|_D \leq \exp\Big(\frac{1}{4\alpha}\, t\Big)\, \|u_\alpha(0)\|_D.$$

These estimate lead us to conclude that, when studying parabolic equations with the use of the quasi-inversion method, solutions of these problems should be sought only for small time intervals.

3.3. SIMPLIFICATION OF EQUATION OF THE QUASI-INVERSION METHOD

We analyze the coefficients of equation (3.11). To this end, we write and transform the coefficients A, B, and C of integro-differential equation of the problem (3.7)–(3.9) introducing another notation $f = \varphi_x/\varphi$:

$$A = EJ_0 + vJ_1 = EJ_0 + v\int_0^t (1 - E)J_0\, \mathrm{d}t,$$

$$B = E\left(J_0^2 - f_x - \int_0^t v_{xx}\,\mathrm{dt}\right)$$
$$+ v\left[2J_0 \int_0^t (1-E)J_0\,\mathrm{dt} - \int_0^t (1-E)\left(J_0^2 + f_x + \int_0^t v_{xx}\,\mathrm{dt}\right)\mathrm{dt}\right],$$

$$C = \left(J_0^2 + f_x + \int_0^t v_{xx}\,\mathrm{dt}\right)$$
$$\times \int_0^t (1-E)J_0\,\mathrm{dt} - J_0 \int_0^t (1-E)\left(J_0^2 + f_x + \int_0^t v_{xx}\,\mathrm{dt}\right)\mathrm{dt}.$$

Calculate the derivatives of the functionals E and J_0 entering to these coefficients:

$$E_x = (1-E)(f - J_0), \qquad E_t = -v(1-E),$$

$$E_{xx} = -(1-E)(J_{0x} - f_x + J_0^2 - 2fJ_0 + f^2)$$

$$J_{0x} = \int_0^t v_{xx}\,\mathrm{dt} + f_x, \qquad J_{0xx} = \int_0^t v_{xxx}\,\mathrm{dt} + f_{xx}, \qquad J_{0t} = v_x.$$

Calculate the derivatives of the coefficients A, B, and C entering to the coefficients of equation (3.11)

$$A_x = (1-E)(f - J_0)J_0 + EJ_{0x} + v_x \int_0^t (1-E)J_0\,\mathrm{dt}$$
$$+ v \int_0^t (1-E)(J_{0x} - J_0 f + J_0^2)\,\mathrm{dt},$$

$$A_{xx} = (1-E)(-3J_0J_{0x} + f_xJ_0 - J_0^3 + 2fJ_0^2 - f^2J_0 + 2fJ_{0x} + EJ_{0xx}$$
$$+ EJ_{0xx} + v_{xx} \int_0^t (1-E)J_0\,\mathrm{dt} + 2v_x \int_0^t (1-E)(J_{0x} - fJ_0 + J_0^2)\,\mathrm{dt}$$
$$+ v \int_0^t (1-E)(3J_0J_{0x} - f_xJ_0 + J_0^3 - 2fJ_0^2 + f^2J_0 - 2fJ_{0x} + J_{0xx})\,\mathrm{dt},$$

$$A_t = Ev_x + v_t \int_0^t (1-E)J_0\,\mathrm{dt},$$

$$\begin{aligned}
B_x &= (1-E)(f-J_0)\left(J_0^2 - f_x - \int_0^t v_{xx}\,\mathrm{d}t\right)\\
&\quad + E\left(2J_0J_{0x} - f_{xx} - \int_0^t v_{xxx}\,\mathrm{d}t\right)\\
&\quad + v_x\left[2J_0\int_0^t (1-E)J_0\,\mathrm{d}t - \int_0^t (1-E)\left(J_0^2 + f_x + \int_0^t v_{xx}\,\mathrm{d}t\right)\mathrm{d}t\right]\\
&\quad + v\left\{2J_{0x}\int_0^t (1-E)J_0\,\mathrm{d}t + 2J_0\int_0^t (1-E)(J_{0x} - fJ_0 + J_0^2)\,\mathrm{d}t\right.\\
&\qquad - \int_0^t (1-E)\left[\left(2J_0J_{0x} + f_{xx} + \int_0^t v_{xxx}\,\mathrm{d}t\right)\right.\\
&\qquad\qquad \left.\left. - (f-J_0)\left(J_0^2 + f_x + \int_0^t v_{xx}\,\mathrm{d}t\right)\right]\mathrm{d}t\right\},
\end{aligned}$$

$$\begin{aligned}
B_{xx} &= \int_0^t (1-E)\left(J_0^2 - f_x - \int_0^t v_{xx}\,\mathrm{d}t\right)[(f_x - J_{0x}) - (f-J_0)^2]\\
&\quad + 2(1-E)(f-J_0)\left(2J_0J_{0x} - f_{xx} - \int_0^t v_{xxx}\,\mathrm{d}t\right)\\
&\quad + E\left(2J_{0x}^2 + 2J_0J_{0x} - f_{xxx} - \int_0^t v_{xxxx}\,\mathrm{d}t\right)\\
&\quad + v_{xx}\left[2J_0\int_0^t (1-E)J_0\,\mathrm{d}t - \int_0^t (1-E)\left(J_0^2 + f_x + \int_0^t v_{xx}\,\mathrm{d}t\right)\mathrm{d}t\right]\\
&\quad + 2v_x\left\{2J_{0x}\int_0^t (1-E)J_0\,\mathrm{d}t + 2J_0\int_0^t (1-E)(J_{0x} - fJ_0 + J_0^2)\,\mathrm{d}t\right.\\
&\qquad - \int_0^t (1-E)\left[\left(2J_0J_{0x} + f_{xx} + \int_0^t v_{xxx}\,\mathrm{d}t\right)\right.\\
&\qquad\qquad \left.\left. - (f-J_0)\left(J_0^2 + f_x + \int_0^t v_{xx}\,\mathrm{d}t\right)\right]\mathrm{d}t\right\}\\
&\quad + v\left\{2J_{0xx}\int_0^t (1-E)J_0\,\mathrm{d}t + 4J_{0x}\int_0^t (1-E)(J_{0x} - fJ_0 + J_0^2)\,\mathrm{d}t\right.\\
&\qquad + 2J_0\int_0^t (1-E)[(f-J_0)^2J_0 - (f_x - J_{0x})J_0\\
&\qquad\qquad - 2(f-J_0)J_{0x} + J_{0xx}]\,\mathrm{d}t\\
&\qquad - \int_0^t (1-E)\left[(f-J_0)^2\left((J_0^2 + f_x + \int_0^t v_{xx}\,\mathrm{d}t\right)\right.\\
&\qquad\qquad - (f_x - J_{0x})\left(J_0^2 + f_x + \int_0^t v_{xx}\,\mathrm{d}t\right)
\end{aligned}$$

$$- 2(f - J_0)\Big(2J_0J_{0x} + f_{xx} + \int_0^t v_{xxx}\,\mathrm{dt}\Big)$$
$$+ \Big(2J_{0x}^2 + 2J_0J_{0xx} + f_{xxx} + \int_0^t v_{xxxx}\,\mathrm{dt}\Big)\Big]\Big\},$$

$$B_t = -Ev_{xx} + 2v_x\Big[EJ_0 + v\int_0^t (1-E)J_0\,\mathrm{dt}\Big]$$
$$+ v_t\Big[2J_0\int_0^t (1-E)J_0\,\mathrm{dt} - \int_0^t (1-E)\Big(J_0^2 + f_x + \int_0^t v_{xx}\,\mathrm{dt}\Big)\,\mathrm{dt}\Big],$$

$$C_x = \Big(2J_0J_{0x} + f_{xx} + \int_0^t v_{xxx}\,\mathrm{dt}\Big)\int_0^t (1-E)J_0\,\mathrm{dt}$$
$$+ \Big(J_0^2 + f_x + \int_0^t v_{xx}\,\mathrm{dt}\Big)\int_0^t (-E_xJ_0 + (1-E)J_{0x})\,\mathrm{dt}$$
$$- J_{0x}\int_0^t (1-E)\Big(J_0^2 + f_x + \int_0^t v_{xx}\,\mathrm{dt}\Big)$$
$$- J_0\int_0^t \Big[- E_x\Big(J_0^2 + f_x + \int_0^t v_{xx}\,\mathrm{dt}\Big)$$
$$+ (1-E)\Big(2J_0J_{0x} + f_{xx} + \int_0^t v_{xxx}\,\mathrm{dt}\Big)\Big]\,\mathrm{dt},$$

$$C_{xx} = \Big(2J_{0x}^2 + 2J_0J_{0xx} + f_{xxx} + \int_0^t v_{xxxx}\,\mathrm{dt}\Big)\int_0^t (1-E)J_0\,\mathrm{dt}$$
$$+ 2\Big(2J_0J_{0x} + f_{xx} + \int_0^t v_{xxx}\,\mathrm{dt}\Big)\int_0^t [-E_xJ_0 + (1-E)J_{0x}]\,\mathrm{dt}$$
$$+ \Big(J_0^2 + f_x + \int_0^t v_{xx}\,\mathrm{dt}\Big)\int_0^t [-E_{xx}J_0 - 2E_xJ_{0x} + (1-E)J_{0xx}]\,\mathrm{dt}$$
$$- J_{0xx}\int_0^t (1-E)\Big(J_0^2 + f_x + \int_0^t v_{xx}\,\mathrm{dt}\Big)$$
$$- 2J_{0x}\int_0^t \Big[- E_x\Big(J_0^2 + f_x + \int_0^t v_{xx}\,\mathrm{dt}\Big)$$
$$+ (1-E)\Big(2J_0J_{0x} + f_{xx} + \int_0^t v_{xxx}\,\mathrm{dt}\Big)\Big]\,\mathrm{dt}$$
$$- J_0\int_0^t \Big[- E_{xx}\Big(J_0^2 + f_x + \int_0^t v_{xx}\,\mathrm{dt}\Big)$$

$$- 2E_x\left(2J_0J_{0x} + f_{xx} + \int_0^t v_{xxx}\,\mathrm{d}t\right)$$

$$+ (1 - E)\left(2J_{0x}^2 + 2J_0J_{0xx} + f_{xxx} + \int_0^t v_{xxxx}\,\mathrm{d}t\right)\Bigg]\,\mathrm{d}t,$$

$$C_t = (2J_0J_{0t} + v_{xx})\int_0^t (1 - E)J_0\,\mathrm{d}t + \left(J_0^2 + f_x + \int_0^t v_{xx}\,\mathrm{d}t\right)(1 - E)J_0$$

$$- J_{0t}\int_0^t (1 - E)\left(J_0^2 + f_x + \int_0^t v_{xx}\,\mathrm{d}t\right)\mathrm{d}t$$

$$- J_0(1 - E)\left(J_0^2 + f_x + \int_0^t v_{xx}\,\mathrm{d}t\right).$$

The obtained expressions show that the coefficients of equation (3.11) are very unwieldy. Consider the ways of their simplification. As it was noted in Section 3.2,when solving problems for parabolic equations by means of the quasi-inversion method, the time interval for the sought solution should be small. Thus, the integrals entering the coefficients of equation (3.11) have small upper limit t. To simplify the coefficients, we calculate the involved integrals using the mean-value theorem, which in the case of the composite integrand function $F(v(x,t),t)$ takes the form

$$\int_0^t F(v(x,t),t)\,\mathrm{d}t = F(v(x,\bar{t}),\bar{t})\,t, \qquad 0 < \bar{t} < t_1.$$

Furthermore, writing the expressions for E and the corresponding derivatives, we use the power series expansion of the exponent in powers of t:

$$E = 1 - \exp\left(\int_0^t v(x,t)\,\mathrm{d}t\right) = -\bar{v}t - \frac{1}{2}(\bar{v}t)^2 + O(t^3),$$

where $\bar{v} = v(x,\bar{t})$. As a result, we obtain the following relations:

$$J_0 = \bar{v}_x t + f, \qquad J_{0x} = \bar{v}_{xx}t + f_x, \qquad J_{0xx} = \bar{v}_{xxx}t + f_{xx}, \qquad J_{0t} = \bar{v}_x,$$

$$J_1 = ft + \frac{1}{2}(\bar{v}_x + f\bar{v})t^2, \qquad J_{1x} = f_x t + \frac{1}{2}(\bar{v}_{xx} + f_x\bar{v} + f\bar{v}_x)t^2,$$

$$J_{1xx} = f_{xx}t + \frac{1}{2}(\bar{v}_{xxx} + f_{xx}\bar{v} + 2f_x\bar{v}_x + f\bar{v}_{xx})t^2, \qquad J_{1t} = f + (\bar{v}_x t + f)t,$$

$$E = -\bar{v}t - \frac{1}{2}\bar{v}^2t^2, \qquad E_x = -\bar{v}_x t - \bar{v}\bar{v}_x t^2,$$

$$E_{xx} = -\bar{v}_{xx}t - (\bar{v}_x^2 + \bar{v}\bar{v}_{xx})t^2, \qquad E_t = -\bar{v} - \bar{v}^2t - \frac{1}{2}\bar{v}^3t^2,$$

$$J_2 = (f_x + f^2) + (2f\bar{v}_x + \bar{v}_{xx})t + \bar{v}_x^2 t^2,$$

$$J_{2x} = (f_{xx} + 2ff_x) + (2f_x\bar{v}_x + 2f\bar{v}_{xx} + \bar{v}_{xxx})t + 2\bar{v}_x\bar{v}_{xx}t^2,$$

$$J_{2xx} = (f_{xxx} + 2f_x^2 + 2ff_{xx}) + (2f_{xx}\bar{v}_x + 4f_x\bar{v}_{xx} + 2f\bar{v}_{xxx} + \bar{v}_{xxx})t + (2\bar{v}_{xx}^2 + 2\bar{v}_x\bar{v}_{xxx})t^2,$$

$$J_{2t} = (2f\bar{v}_x + \bar{v}_{xx}) + 2\bar{v}_x^2 t,$$

$$J_3 = (f_x + f^2)t + \frac{1}{2}[(f_x + f^2)\bar{v} + (2f\bar{v}_x + \bar{v}_{xx})]t^2,$$

$$J_{3x} = (f_{xx} + 2ff_x)t + \frac{1}{2}[(f_{xx} + 2ff_x)\bar{v} + (f_x + f^2)\bar{v}_x + 2f_x\bar{v}_x + 2f\bar{v}_{xx} + \bar{v}_{xxx}]t^2,$$

$$J_{3xx} = (f_{xxx} + 2f_x^2 + 2ff_{xx})t + \frac{1}{2}[(f_{xxx} + 2f_x^2 + 2ff_{xx})\bar{v} + 2(f_{xx} + 2ff_x)\bar{v}_x + (f_x + f^2)\bar{v}_{xx} + 2(f_{xx}\bar{v}_x + 2f_x\bar{v}_{xx} + f\bar{v}_{xxx}) + \bar{v}_{xxxx}]t^2,$$

$$J_{3t} = (f_x + f^2) + [(f_x + f^2)\bar{v} + (2f\bar{v}_x + \bar{v}_{xx})]t.$$

Deducing these formulas we preserved the terms containing t^2 and threw away the terms of order $O(t^3)$.

Apply the mean-value theorem to the definite integrals in the coefficients of the integro-differential equation and calculate them with the accuracy that can be estimate as follows:

$$\int_0^t f(x,t)\,\mathrm{d}t = f(x,\bar{t})t + O(\Delta f\, t), \qquad 0 < \bar{t} < t,$$

where

$$\Delta f = \left|\max_t f(x,t) - \min_t f(x,t)\right|.$$

We used the above approximate relations for deducing the approximate formulas for the coefficients of equation (3.7) and their derivatives. We preserved the terms containing t^2 and threw away the terms of order $O(t^3)$. When simplifying the expressions, we took into account that the coefficients of equation (3.7) contain the derivatives A_t, B_t, and C_t. To provide the required accuracy of calculation of these derivatives it is necessary to construct the approximate expressions with the accuracy of order up to $O(t^3)$

omitting the terms of order $O(t^4)$. Then we obtain

$$A = -\frac{1}{2}\bar{v}\bar{v}_x t^2 - \frac{1}{2}\bar{v}^2\bar{v}_x t^3, \qquad A_x = -\frac{1}{2}(\bar{v}_x^2 + \bar{v}\bar{v}_{xx})t^2,$$

$$A_{xx} = -\frac{1}{2}(3\bar{v}_x\bar{v}_{xx} + \bar{v}\bar{v}_{xxx})t^2, \qquad A_t = -\bar{v}\bar{v}_x t - \frac{3}{2}\bar{v}^2\bar{v}_x t^2,$$

$$B = \frac{1}{2}\bar{v}\bar{v}_{xx}t^2 + \frac{1}{2}\bar{v}^2\bar{v}_{xx}t^3 - \frac{1}{2}\bar{v}^2\bar{v}_x^2 t^4, \qquad B_x = \frac{1}{2}(\bar{v}_x\bar{v}_{xx} + \bar{v}\bar{v}_{xxx})t^2,$$

$$B_{xx} = \frac{1}{2}(\bar{v}_{xx}^2 + 2\bar{v}\bar{v}_{xxx} + \bar{v}\bar{v}_{xxxx})t^2, \qquad B_t = \bar{v}_x\bar{v}_{xx}t + \frac{3}{2}\bar{v}^2\bar{v}_{xx}t^2,$$

$$C = \frac{1}{2}[f\bar{v}_{xx} + (f^2 - f_x)\bar{v}_x]t^2 + f\bar{v}_x(\bar{v}_x + \frac{1}{2}f\bar{v})t^3,$$

$$C_x = \frac{1}{2}[f\bar{v}_{xxx} + f^2\bar{v}_{xx} + (2ff_x - f_{xx})\bar{v}_x]t^2,$$

$$\begin{aligned} C_{xx} = \frac{1}{2}[&(f_x + f^2)\bar{v}_{xxx} + (4ff_x - f_{xx})\bar{v}_{xx} \\ &+ (2ff_{xx} + 2f_x^2 - f_{xxx})\bar{v}_x + f\bar{v}_{xxxx}]t^2, \end{aligned}$$

$$C_t = [f\bar{v}_{xx} + (f^2 - f_x)\bar{v}_x]t + 3f\bar{v}_x(\bar{v}_x + \frac{1}{2}f\bar{v})t^2.$$

Estimate the order of smallness for each term in the equation of quasi-inversion method. For this purpose we multiply this equation by ε_1 and then rewrite it as follows:

$$\begin{aligned} &A^2\tilde{v}_{xxxx} + 4AA_x\tilde{v}_{xxx} \\ &+ [2A_x^2 + 2AA_{xx} - B^2 + A_xB + AB_x + (A_tC + AC_t) - \varepsilon_1 A]\tilde{v}_{xx} \\ &+ [-2BB_x + A_{xx}B + 2A_xB_x + AB_{xx} + (B_tC + BC_t) - \varepsilon_1 B]\tilde{v}_x \\ &+ [-A_{xx}C - 2A_xC_x - AC_{xx} + B_xC + BC_x - 2CC_t + \varepsilon_1 C]\tilde{v}_t \\ &- (C^2 + \varepsilon_1\varepsilon_2)\tilde{v}_{tt} + 2(BC - A_xC - AC_x)\tilde{v}_{xt} = 0. \end{aligned}$$

Consider the coefficients of derivatives $\tilde{v}_{xx}$ and $\tilde{v}_x$ in this equation. Taking into account the smallness of the time interval over which this equation is integrated we make the following assumption. Let all mean values $\bar{v}$, $\bar{v}_x$, and $\bar{v}_{xx}$ introduced above be the same for all integrals (further this assumption will be substantiated). Furthermore, assume that $\bar{v}_x = \tilde{v}_x$ and

$\bar{v}_{xx} = \tilde{v}_{xx}$. Construct the expression $A\tilde{v}_{xx} + B\tilde{v}_x$. Basing on these assumptions we obtain

$$\begin{aligned} A\tilde{v}_{xx} + B\tilde{v}_x &= -\frac{1}{2}\,\bar{v}\bar{v}_x\tilde{v}_{xx}t^2 - \frac{1}{2}\,\bar{v}^2\bar{v}_x\tilde{v}_{xx}t^3 + \frac{1}{2}\,\bar{v}\bar{v}_{xx}\tilde{v}_x t^2 \\ &\quad + \frac{1}{2}\,\bar{v}^2\bar{v}_{xx}\tilde{v}_x t^3 - \frac{1}{2}\,\bar{v}^2\bar{v}_x^2\tilde{v}_x t^4 \\ &\simeq -\frac{1}{2}\,v^2 v_x^3 t^4. \end{aligned}$$

The upper accent of v in the right-hand side of the equation are omitted. By virtue of the above assumptions, we can write

$$A\tilde{v}_{xx} + B\tilde{v}_x \simeq -\frac{1}{2}\,\tilde{v}^2\tilde{v}_x^3 t^4.$$

Now we use the fact that the solution of the quasi-inversion problem is an approximate solution of the problem (3.7)–(3.9): $\tilde{v} \approx v$. In this case equation (3.7) leads to the equality which takes into account our done assumption:

$$A\tilde{v}_{xx} + B\tilde{v}_x \simeq C\tilde{v}_t$$

$$\Downarrow$$

$$-\frac{1}{2}\,\tilde{v}^2\tilde{v}_x^3 t^4 \simeq \frac{1}{2}[f\tilde{v}_{xx} + (f^2 - f_x)\tilde{v}_x]t^2\tilde{v}_t.$$

From this equality it follows that our assumptions and simplifications are valid when $\tilde{v}_t = O(t^2)$. Thus, for further study of equation (3.7) we assume $\tilde{v}_t = O(t^2)$. According to the above, $\tilde{v}_{tt} = O(t)$ and $\tilde{v}_{xt} = O(t^2)$.

Return now to the study of the coefficients of equation (3.7). Write the derivatives $\tilde{v}_x$ and $\tilde{v}_{xx}$ appearing in (3.7) together with their coefficients:

$$\begin{aligned} &[(2A_x^2 + 2AA_{xx} - B^2 + A_xB + AB_x) + (A_tC + AC_t) - \varepsilon_1 A]\tilde{v}_{xx} \\ &+ [(-2BB_x + A_{xx}B + 2A_xB_x + AB_{xx}) + (B_tC + BC_t) - \varepsilon_1 B]\tilde{v}_x \\ &\quad = (2A_x^2 + AA_{xx} - B^2 + A_xB + AB_x + A_tC)\tilde{v}_{xx} \\ &\qquad + (-2BB_x + 2A_xB_x + AB_{xx} + AB_{xx} + B_tC)\tilde{v}_x \\ &\qquad + (A_{xx} + C_t - \varepsilon_1)(A\tilde{v}_{xx} + B\tilde{v}_x) \\ &\quad = (2A_x^2 + AA_{xx} - B^2 + A_xB + AB_x)\tilde{v}_{xx} \\ &\qquad + (-2BB_x + 2A_xB_x + AB_{xx})\tilde{v}_x + C(A_t\tilde{v}_{xx} + B_t\tilde{v}_x) \\ &\qquad + (A_{xx} + C_t - \varepsilon_1)(A\tilde{v}_{xx} + B\tilde{v}_x). \end{aligned}$$

By analogy with the previous operations, estimate the expression $A_t\tilde{v}_{xx} + B_t\tilde{v}_x$ using the same assumptions:

$$\begin{aligned} A_t\tilde{v}_{xx} + B_t\tilde{v}_x &= -\bar{v}\bar{v}_x\tilde{v}_{xx}t - \frac{3}{2}\bar{v}^2\bar{v}_x\tilde{v}_{xx}t + \bar{v}\bar{v}_{xx}\tilde{v}_x t \\ &\quad + \frac{3}{2}\bar{v}^2\bar{v}_{xx}\tilde{v}_x t^2 - 2\bar{v}^2\bar{v}_x^2\tilde{v}_{xx}t^3 \\ &= -2\bar{v}^2\bar{v}_x^2\tilde{v}_{xx}t^3. \end{aligned}$$

As a result, we obtain

$$C(A_t\tilde{v}_{xx} + B_t\tilde{v}_x) = O(t^5), \qquad A_{xx}(A_t\tilde{v}_{xx} + B\tilde{v}_x) = O(t^6),$$
$$C_t(A_t\tilde{v}_{xx} + B\tilde{v}_x) = O(t^5).$$

The order of smallness of the term $-\varepsilon_1(A\tilde{v}_{xx} + B\tilde{v}_x)$ containing the regularization parameter ε_1 is determined by the dependence between ε_1 and t. Since the value of the regularization parameter ε_1 will be chosen in the process of calculations, we shall not impose any restrictions upon the order of smallness of this term beforehand. It is obvious that the other coefficients in the considered part of equation (3.11) have the order of smallness $O(t^4)$.

Thus, we can group the terms in the considered part of the equation taking into account the orders of smallness of the involved coefficients:

$$\begin{aligned} &(2A_x^2 + AA_{xx} - B^2 + A_xB + AB_x)\tilde{v}_{xx} \\ &+ (-2BB_x + 2A_xB_x + AB_{xx})\tilde{v}_x + C(A_t\tilde{v}_{xx} + B_t\tilde{v}_x) \\ &+ C_t(A\tilde{v}_{xx} + B\tilde{v}_x) + A_{xx}(A\tilde{v}_{xx} + B\tilde{v}_x) - \varepsilon_1(A\tilde{v}_{xx} + B\tilde{v}_x). \end{aligned}$$

We now turn to the other terms of equation (3.11).

The dependence of the function $\tilde{v}(x,t)$ on t is strictly specified by equation (3.11). However, recalling the way it was introduced, since $u(x,t)$ varies slowly with time in many applications, we can assume that $\tilde{v}(x,t)$ also varies slowly for small values of t. Thus, the order of the terms entering equation (3.11) is determined by the orders of smallness of its coefficients and derivatives $\tilde{v}_t$, $\tilde{v}_{tt}$ and $\tilde{v}_{xt}$. Now we consider other terms of the equation.

First, we consider the term

$$(-A_{xx}C - 2A_xC_x - AC_{xx} + B_xC + BC_x - 2CC_t + \varepsilon_1C)\tilde{v}_t.$$

Taking into account the order of smallness of the derivative $\tilde{v}_t = O(t^2)$ and using the estimates of the orders of smallness for the coefficients A, B and C and their derivatives, we get

$$(-A_{xx}C - 2A_xC_x - AC_{xx} + B_xC + BC_x)\tilde{v}_t = O(t^6), \qquad -2CC_t\tilde{v}_t = O(t^5).$$

The order of smallness of the quantity $\varepsilon_1 C\tilde{v}_t$ depends on the choice of the regularization parameter ε_1. We do not impose any restrictions on the value of this term lest the choice of ε_1 be restricted in the beginning.

Consider the next term $-(C^2 + \varepsilon_1\varepsilon_2)\tilde{v}_{tt}$ in the equation. Its order of smallness can be easily estimated. Indeed, $\tilde{v}_{tt} = O(t)$ and $C^2\tilde{v}_{tt} = O(t^5)$. As before, we do not impose restrictions on the value of $\varepsilon_1\varepsilon_2\tilde{v}_{tt}$.

Finally, we estimate the last term in the equation:

$$2(BC - A_xC - AC_x)\tilde{v}_{xt} = O(t^6),$$

since $\tilde{v}_{tx} = O(t^2)$.

We now write equation (3.11) with substituted approximate values of the coefficients A, B and C and their derivatives. First, consider the following part of the equation:

$$\begin{aligned}
&A^2\tilde{v}_{xxxx} + 4AA_x\tilde{v}_{xxx} + (2A_x^2 + AA_{xx} + A_xB - B^2 + AB_x - \bar{\varepsilon}_1 A)\tilde{v}_{xx} \\
&+ (2A_xB_x - 2BB_x + AB_{xx} - \bar{\varepsilon}_1 B)\tilde{v}_x \\
&= \frac{1}{4}\bar{v}^2\bar{v}_x^2\tilde{v}_{xxxx}t^4 + \bar{v}\bar{v}_x(\bar{v}_x^2 + \bar{v}\bar{v}_{xx})\tilde{v}_{xxx}t^4 \\
&\quad + \frac{1}{2}\bar{v}_x\Big(\bar{v}_x^3 + \frac{5}{2}\bar{v}\bar{v}_x\bar{v}_{xx} + \bar{\varepsilon}_1\bar{v}\Big)\tilde{v}_{xx}t^4 - \frac{1}{2}\bar{v}_{xx}\Big(\bar{v}_x^3 + \frac{5}{2}\bar{v}\bar{v}_x\bar{v}_{xx} + \bar{\varepsilon}_1\bar{v}\Big)\tilde{v}_{xx}t^4 \\
&\quad - \bar{v}\Big[\bar{v}_{xxx}(\bar{v}_x^2 + \bar{v}\bar{v}_{xx}) + \frac{1}{4}\bar{v}\bar{v}_x\bar{v}_{xxxx}\Big]\tilde{v}_xt^4,
\end{aligned} \tag{3.12}$$

where $\bar{\varepsilon}_1 = \varepsilon_1/t$. The assumption $\tilde{v}(x,t) = \bar{v}(x)$, which we used before, leads to the vanishing of the foregoing expression. As a result, equation (3.11), which is the main equation used for calculations, is simplified.

We now estimate the accuracy of the assumption made. We use the estimate $\tilde{v}_t(x,t) = O(t^2)$ obtained before. Let $\bar{v}(x) = v(x,\bar{t})$ and $\tau = t - \bar{t}$. Write the derivative $\tilde{v}_t(x,t)$ as a finite difference using the difference scheme of the first order of accuracy:

$$v_t(x,t) = \frac{1}{\tau}(\tilde{v}(x,t) - v(x,t-\tau)) + O(\tau),$$

where $v(x,t-\tau) = v(x,\bar{t}) = \bar{v}(x)$. Then we obtain

$$\tilde{v}(x,t) - \bar{v}(x) = \tau\tilde{v}_t(x,t) + O(\tau^2), \tag{3.13}$$

$$\begin{aligned}
O(|\tilde{v}(x,t) - \bar{v}(x)|) &= \tau O(t^2) + O(\tau^2) = (t-\bar{t})O(t^2) + O((t-\bar{t})^2) \\
&= O((t-\bar{t})t^2) + O((t-\bar{t})^2) = O((t-\bar{t})t^2 + (t-\bar{t})^2).
\end{aligned}$$

This estimate can be rewritten in the following form:

$$\tilde{v}(x,t) = \bar{v}(x) + O(\tau t^2 + \tau^2),$$

which implies that the error caused by the substitution in the considered part of the equation is $O(\tau t^6 + \tau^2 t^4)$, taking into account the order of smallness of the coefficients in this part of the equation, which is equal to $O(t^4)$.

Thus, after the transformations made, the equation of the quasi-inversion method takes the form

$$\begin{aligned}&C(A_t\tilde{v}_{xx} + B_t\tilde{v}_x) + (C_t + A_{xx})(A\tilde{v}_{xx} + B\tilde{v}_x)\\&+ (-A_{xx}C + B_xC - 2A_xC_x + BC_x - AC_{xx} - 2CC_t + \varepsilon_1 C)\tilde{v}_t\\&- (C^2 + \varepsilon_1\varepsilon_2)\tilde{v}_{tt} + 2(BC - A_xC - AC_x)\tilde{v}_{xt} + O(\tau t^6 + \tau^2 t^4) = 0.\end{aligned}$$

We rewrite the obtained equation preserving the terms with order of smallness $O(t^5)$ and omitting the terms of higher orders of smallness:

$$\begin{aligned}&C(A_t\tilde{v}_{xx} + B_t\tilde{v}_x) + C_t(A\tilde{v}_{xx} + B\tilde{v}_x) + C(\varepsilon_1 - 2C_t)\tilde{v}_t\\&- (C^2 + \varepsilon_1\varepsilon_2)\tilde{v}_{tt} + O(t^6 + \tau t^6 + \tau^2 t^4) = 0.\end{aligned} \tag{3.14}$$

Using the above assumption $\tilde{v}(x,t) = \bar{v}(x)$ for the first two terms of equation (3.14) we can simplify it further. According to the previous calculations, we have

$$A_t\tilde{v}_{xx} + B_t\tilde{v}_x \simeq -2v^2v_x^3t^3, \qquad A\tilde{v}_{xx} + B\tilde{v}_x \simeq -\frac{1}{2}v^2v_x^3t^4.$$

We substitute these expressions into equation (3.14) together with the values obtained for the coefficient C and its derivative C_t

$$(F^2 + 4\tilde{\varepsilon}_1\tilde{\varepsilon}_2)t\tilde{v}_{tt} + 2(2F - \tilde{\varepsilon}_1)F\tilde{v}_t + 6Fv^2v_x^3t^2 + O(t^3 + \tau t + \tau^2 t) = 0,$$

where $F = f\tilde{v}_{xx} + (f^2 - f_x)\tilde{v}_x$, $\tilde{\varepsilon}_1 = \varepsilon_1/t$ and $\tilde{\varepsilon}_2 = \varepsilon_2/t^3$. In the last term of the foregoing equation we omit the upper accents, as was done before. Preserving the same accuracy as for equation (3.14), we can rewrite it in the form

$$(F^2 + 4\tilde{\varepsilon}_1\tilde{\varepsilon}_2)t\tilde{v}_{tt} + (2F - \tilde{\varepsilon}_1)2F\tilde{v}_t + 6F\tilde{v}^2\tilde{v}_x^3t^2 = 0, \tag{3.15}$$

setting $F = f\tilde{v}_{xx} + (f^2 - f_x)\tilde{v}_x$ and thus considering (3.15) as a non-linear partial differential equation of the second order.

Rigorous substantiation of the simplifications made is reduced to the study of the coefficient stability of a solution to the nonlinear partial differential equation of the fourth order (3.11). Such a study is interesting in itself from the mathematical point of view. Here we restrict ourselves to confirming the results of computational experiment (see Samarskii, 1979; Popov and Samarskii, 1988).

In the process of the numerical solution of an initial-boundary value problem for equation (3.11) it is always necessary to carry out approximate calculation of the integrals entering its coefficients. The error in these calculations depend on the chosen method. Whatever it may appear, this error will affect the solution of the equation. This influence will be weak in case of the coefficient stability of the solution. Then we can state that the error in the solution depends upon the error in the calculation of integrals in the coefficients. The estimate of this dependence will serve as criterion for the choice of the method of approximate calculation of the integrals. In particular, it will serve as a substantiation of the simplifications made in the calculation of the coefficients that also lead to the simplification of the original equation.

The study of the coefficient stability of solutions to the initial-boundary value problem for a nonlinear partial differential equation of the fourth order with two independent variables and coefficients containing integrals is an independent mathematical problem. There are examples of the study of coefficient stability (co-stability) for some simple differential equations (see Tikhonov and Samarskii, 1961; Samarskii, 1977).

The computational experiment for the investigation of the co-stability of the initial-boundary value problem for equation (3.11) is organized as follows. One should either solve an example problem having exact analytic solution or use different methods of calculating integrals in the coefficients in the process of calculation, every time comparing the results of the numerical solution of equation (3.11). If the solution is co-stable, then the results of the calculations should be close for different calculation methods.

3.4. SIMPLIFICATION OF THE QUASI-INVERSION PROBLEM

In the previous sections we presented the equation of the quasi-inversion method corresponding to the original problem statement. Now we formulate the quasi-inversion problem in the complete form.

Find a solution to the equation

$$\frac{1}{\varepsilon_1} P^* P\tilde{v} + P\tilde{v} - \varepsilon_2 \tilde{v}_{tt} = 0$$

defined on $(x_0, x_1) \times (0, t_1)$ and satisfying the boundary conditions

$$\tilde{v}(x_0, t) = \mu_0(t), \qquad \tilde{v}_x(x_0, t) = \eta_0(t), \tag{3.16}$$

$$\tilde{v}(x_1, t) = \mu_1(t), \qquad \tilde{v}_x(x_1, t) = \eta_1(t), \qquad -\varepsilon < t \le t_1, \tag{3.17}$$

$$\tilde{v}(x, 0) - \varepsilon_2 \tilde{v}_t(x, 0) - \frac{1}{\varepsilon_1} P\tilde{v}(x, 0) = v^*(x), \tag{3.18}$$

$$\varepsilon_2 \tilde{v}_t(x, t_1) + \frac{1}{\varepsilon_1}(P\tilde{v})(x, t_1) = 0, \qquad x_0 \le x \le x_1. \tag{3.19}$$

Here $v^*(x)$ is an arbitrary function. For example, one can assume $v^*(x) = 0$. Examples will be given to show that the results of calculations do not depend substantially on the choice of the function $v^*(x)$.

So, the quasi-inversion problem corresponding to the original problem includes two regularization parameters ε_1 and ε_2.

We shall make simplifying transformations of the boundary conditions analogous to those made in the previous section for the quasi-inversion equation. Consider the boundary condition (3.18), where $P\tilde{v} = -A\tilde{v}_{xx} - B\tilde{v}_x + C\tilde{v}_t$. After the simplifications we get

$$P\tilde{v} \simeq \frac{1}{2} v^2 v_x^3 t^4 + \frac{1}{2} F\tilde{v}_t t^2. \tag{3.20}$$

From (3.20) it follows that $(P\tilde{v})(x, 0) = 0$. Then the boundary (3.18) condition takes the form

$$\tilde{v}(x, 0) - \varepsilon_2 \tilde{v}_t(x, 0) = v^*(x). \tag{3.21}$$

For the boundary condition (3.19) we make the same simplification:

$$(P\tilde{v})(x, t_1) \simeq \frac{1}{2} \tilde{v}^2(x, t_1) \tilde{v}_x^3(x, t_1) t_1^4 + \frac{1}{2} F(x, t_1) \tilde{v}_t(x, t_1) t_1^2. \tag{3.22}$$

As before, in the formula (3.22) we set $\bar{v} = \tilde{v}$. Substituting (3.22) into the boundary condition (3.19) and multiply the latter by $2\varepsilon_1$:

$$(2\varepsilon_1 \varepsilon_2 + F t_1^2) \tilde{v}_t(x, t_1) + \tilde{v}^2(x, t_1) \tilde{v}_x^3(x, t_1) t_1^4 = 0. \tag{3.23}$$

So, we consider equation (3.10) in the rectangle $\{(x_0, x_1) \times (0, t_1]\}$ together with the additional conditions (3.16), (3.16), (3.21) and (3.23). Equation (3.10) written in the form (3.15) after the transformations made in the previous section, together with the mentioned conditions, make up a problem with overdetermined set of boundary conditions. For this reason, we return to the transformations simplifying equation (3.11).

Equation (3.11) contains the fourth-order derivative with respect to x. It was included in the set of summands of small order forming the remainder term of order $O(t^3+\tau t+\tau^2 t)$ in the process of simplification and calculating the orders of smallness of the equation terms. It was the reason for which the order of the equation changed and the boundary conditions were overdetermined. To avoid this situation, we should take into account the terms containing $\tilde{v}_{xxxx}$ in the equation. First, we simplify the formula (3.12):

$$\begin{aligned}
&A^2\tilde{v}_{xxxx} + 4AA_x\tilde{v}_{xxx} + (2A_x^2 + AA_{xx} + A_xB - B^2 + AB_x - \varepsilon_1 A)\tilde{v}_{xx}\\
&+ (2A_xB_x - 2BB_x + AB_{xx} - \varepsilon_1 B)\tilde{v}_x\\
&\quad = \frac{1}{4}\bar{v}^2\bar{v}_x(\bar{v}_x\tilde{v}_{xxxx} - \tilde{v}\bar{v}_{xxxx})t^4\\
&\qquad + [\bar{v}\bar{v}_x^2(\bar{v}_x\tilde{v}_{xxx} - \tilde{v}_x\bar{v}_{xxx}) + \bar{v}\bar{v}_{xx}(\bar{v}_x\tilde{v}_{xxx} - \tilde{v}_x\bar{v}_{xxx})]t^4\\
&\qquad + \frac{1}{2}\Big(\bar{v}_x^3 + \frac{5}{2}\bar{v}\bar{v}_x\bar{v}_{xx} + \varepsilon_1\bar{v}\Big)(\bar{v}\tilde{v}_{xx} - \tilde{v}\bar{v}_{xx})t^4
\end{aligned} \tag{3.24}$$

Using the formula (3.13), we transform the differences that enter the obtained result (3.24)

$$\bar{v}(x) = v(x,\bar{t}) = \tilde{v}(x,t) - \tau_1\tilde{v}_t(x,t) + O(\tau_1^2), \qquad \tau_1 = t - \bar{t},$$

$$\bar{v}_{x^i} = \tilde{v}_{x^i} - \tau_1\tilde{v}_{tx^i} + O(\tau_1^2), \qquad i = 1,2,3,4.$$

$$\begin{aligned}
\bar{v}_x\tilde{v}_{xx} - \tilde{v}_x\bar{v}_{xx} &= (\tilde{v}_x - \tau_1\tilde{v}_{tx})\tilde{v}_{xx} - \tilde{v}_x(\tilde{v}_{xx} - \tau_1\tilde{v}_{tx^2})\\
&= \tau_1(\tilde{v}_x\tilde{v}_{txx} - \tilde{v}_{tx}\tilde{v}_{xx} = \tau_1\Big(\frac{\tilde{v}_{xxt}}{\tilde{v}_{xx}} - \frac{\tilde{v}_{xt}}{\tilde{v}_x}\Big)\tilde{v}_x\tilde{v}_{xx},
\end{aligned}$$

$$\bar{v}_x\tilde{v}_{xxx} - \tilde{v}_x\bar{v}_{xxx} = \tau_1\Big(\frac{\tilde{v}_{xxxt}}{\tilde{v}_{xxx}} - \frac{\tilde{v}_{xt}}{\tilde{v}_x}\Big)\tilde{v}_x\tilde{v}_{xxx},$$

$$\bar{v}_x\tilde{v}_{xxxx} - \tilde{v}_x\bar{v}_{xxxx} = \tau_1\Big(\frac{\tilde{v}_{xxxxt}}{\tilde{v}_{xxxx}} - \frac{\tilde{v}_{xt}}{\tilde{v}_x}\Big)\tilde{v}_x\tilde{v}_{xxxx}.$$

Then the following equalities are valid: (a) for any continuously differentiable function $y(x)$

$$y(x) = |y(x)|\operatorname{sign}((y(x)), \qquad y'(x) = (|y(x)|)'\operatorname{sign}(y(x)),$$

$$\frac{y'(x)}{y(x)} = \frac{(|y(x)|)'\operatorname{sign}(y(x))}{|y(x)|\operatorname{sign}(y(x))} = \frac{(|y(x)|)'}{|y(x)|} = (\ln|y(x)|)'_x.$$

$$\begin{aligned}
\bar{v}_x\tilde{v}_{xx} - \tilde{v}_x\bar{v}_{xx} &= \tau_1\Big(\frac{\tilde{v}_{xxt}}{\tilde{v}_{xx}} - \frac{\tilde{v}_{xt}}{\tilde{v}_x}\Big)\tilde{v}_x\tilde{v}_{xx}\\
&= [(\ln|\tilde{v}_{xx}|)'_t - (\ln|\tilde{v}_x|)'_t]\tilde{v}_x\tilde{v}_{xx}\tau_1 = \Big(\ln\Big|\frac{\tilde{v}_{xx}}{\tilde{v}_x}\Big|\Big)'_t\tilde{v}_x\tilde{v}_{xx}\tau_1,
\end{aligned}$$

$$\bar{v}_x\tilde{v}_{xxx} - \tilde{v}_x\bar{v}_{xxx} = \left(\ln\left|\frac{\tilde{v}_{xxx}}{\tilde{v}_x}\right|\right)'_t\bar{v}_x\tilde{v}_{xxx}\tau_1,$$

$$\begin{aligned}\bar{v}_x\tilde{v}_{xxxx} - \tilde{v}_x\bar{v}_{xxxx} &= \left(\ln\left|\frac{\tilde{v}_{xxxx}}{\tilde{v}_x}\right|\right)'_t\bar{v}_x\tilde{v}_{xxxx}\tau_1\\ &= \left(\ln\left|\frac{\tilde{v}_{xx}}{\tilde{v}_x}\right| + \ln\left|\frac{\tilde{v}_{xxx}}{\tilde{v}_{xx}}\right| + \ln\left|\frac{\tilde{v}_{xxxx}}{\tilde{v}_{xxx}}\right|\right)'_t\bar{v}_x\tilde{v}_{xxxx}\tau_1\\ &= (\ln|(\ln|\tilde{v}_x|)_x(\ln|\tilde{v}_{xx}|)_x(\ln|\tilde{v}_{xxx}|)_x|)_t\bar{v}_x\tilde{v}_{xxxx}\tau_1.\end{aligned}$$

(b) for the case under consideration. We now make the analogous transformations:

$$\begin{aligned}\bar{v}\bar{v}_x^2 + \bar{v}^2\bar{v}_x &= \bar{v}^2\bar{v}_x\left(\frac{\bar{v}_x}{\bar{v}} + \frac{\bar{v}_{xx}}{\bar{v}_x}\right) = \bar{v}^2\bar{v}_x(\ln|\bar{v}\bar{v}_x|)'_x\\ &= \frac{1}{2}\bar{v}^2\bar{v}_x(\ln|(\bar{v}^2)'_x|)'_x = \frac{1}{2}\bar{v}^2\bar{v}_x\frac{(\bar{v}^2)''_{xx}}{(\bar{v}^2)'_x} = \frac{1}{4}\bar{v}(\bar{v}^2)''_{xx},\end{aligned}$$

$$\bar{v}_x^3 + \frac{5}{2}\bar{v}\bar{v}_x\bar{v}_{xx} + \varepsilon_1\bar{v} = \bar{v}\bar{v}_x^2\left(\frac{\bar{v}_x}{\bar{v}} + \frac{5}{2}\frac{\bar{v}_{xx}}{\bar{v}_x}\right) + \varepsilon_1\bar{v} = \bar{v}(\bar{v}_x^2(\ln|\bar{v}\bar{v}_x^2\sqrt{\bar{v}_x}|)'_x + \varepsilon_1).$$

Rewrite the formula (3.12) taking into account the transformations made:

$$\begin{aligned}&A^2\tilde{v}_{xxxx} + 4AA_x\tilde{v}_{xxx} + (2A_x^2 + AA_{xx} + A_xB - B^2 + AB_x - \varepsilon_1A)\tilde{v}_{xx}\\ &+ (2A_xB_x - 2BB_x + AB_{xx} - \varepsilon_1B)\tilde{v}_x\\ &= \frac{1}{4}\left(\ln\left|\frac{\tilde{v}_{xxxx}}{\tilde{v}_x}\right|\right)'_t\bar{v}^2\bar{v}_x\bar{v}_x\tilde{v}_{xxxx}\tau_1t^4 + (2A_xB_x - 2BB_x + AB_{xx} - \varepsilon_1B)\tilde{v}_x\\ &= \frac{1}{4}\left(\ln\left|\frac{\tilde{v}_{xxxx}}{\tilde{v}_x}\right|\right)'_t\bar{v}^2\bar{v}_x\bar{v}_x\tilde{v}_{xxxx}\tau_1t^4 + \frac{1}{4}\left(\ln\left|\frac{\tilde{v}_{xxx}}{\tilde{v}_x}\right|\right)'_t\bar{v}(\bar{v}^2)_{xx}\bar{v}_x\tilde{v}_{xxx}\tau_1t^4\\ &+ \frac{1}{2}\left(\ln\left|\frac{\tilde{v}_{xx}}{\tilde{v}_x}\right|\right)'_t\left(\bar{v}_x^3 + \frac{5}{2}\bar{v}\bar{v}_x\bar{v}_{xx} + \varepsilon_1\bar{v}\right)\bar{v}_x\tilde{v}_{xx}\tau_1t^4\end{aligned}$$

All the terms in the obtained expression have the same order of smallness. We drop the second and third terms while preserving the first one, for if we drop the latter, the order of the equation will change. As follows from the specified form of the boundary conditions, this will cause the problem to become conditionally well-posed again, since the boundary conditions will again become overdetermined. It is convenient to consider the preserved first term in the following form:

$$\begin{aligned}\frac{1}{4}\bar{v}^2\bar{v}_x(\bar{v}_x\tilde{v}_{xxxx} - \tilde{v}_x\bar{v}_{xxxx})t^4 &= \frac{1}{4}\tilde{v}^2\bar{v}_x(\tilde{v}_x\tilde{v}_{xxxxt} - \tilde{v}_{xt}\tilde{v}_{xxxx})\tau_1t^4 + O(\tau_1^2t^4)\\ &= \left(-\frac{1}{4}\tilde{v}^2\bar{v}_x\tilde{v}_{xt}\tau_1t^4\right)\tilde{v}_{xxxx} + \left(\frac{1}{4}\tilde{v}^2\bar{v}_x\tilde{v}_{xxxxt}\tau_1t^4\right))\tilde{v}_x + O(\tau_1^2t^4).\end{aligned}$$

On making the simplifications, we write the equation of the quasi-inversion method for the coefficient inverse problem under consideration:

$$(-\tilde{v}^2\tilde{v}_x\tilde{v}_{xt}\tau_1 t^4)\tilde{v}_{xxxx} + (\tilde{v}^2\tilde{v}_x\tilde{v}_{xxxxt}\tau_1 t^4)\tilde{v}_x + (F^2 + 4\tilde{\varepsilon}_1\tilde{\varepsilon}_2)t\tilde{v}_{tt}$$
$$+ (2F - \tilde{\varepsilon}_1)2F\tilde{v}_t + 6Fv^2v_x^3t^2 + O(t^3 + \tau t + \tau^2 t) = 0.$$

STATEMENT OF THE MODIFIED QUASI-INVERSION PROBLEM

On the set $(x_0; x_1) \times (0; t_1]$, find a solution to the following equation with the corresponding set of additional conditions:

$$-(\tilde{v}^2\tilde{v}_x\tilde{v}_{xt}\tau_1 t)\tilde{v}_{xxxx} + (\tilde{v}^2\tilde{v}_x\tilde{v}_{xxxxt}\tau_1 t)\tilde{v}_x + (F^2 + 4\tilde{\varepsilon}_1\tilde{\varepsilon}_2)t\tilde{v}_{tt}$$
$$+ (2F - \tilde{\varepsilon}_1)2F\tilde{v}_t + 6F\tilde{v}^2\tilde{v}_x^3t^2 = 0;$$

$$\tilde{v}(x_0, t) = \mu_0(t), \qquad \tilde{v}_x(x_0, t) = \eta_0(t),$$
$$\tilde{v}(x_1, t) = \mu_1(t), \qquad \tilde{v}_x(x_1, t) = \eta_1(t), \qquad -\varepsilon < t \le t_1;$$
$$\tilde{v}(x, 0) - \varepsilon_2\tilde{v}_t(x, 0) = \tilde{v}^*(x), \qquad x_0 \le x \le x_1;$$
$$(2\varepsilon_1\varepsilon_2 + Ft_1^2\tilde{v}_t(x, t_1) + \tilde{v}^2(x, t_1)\tilde{v}_x^3(x, t_1)t_1^4 = 0, \qquad x_0 \le x \le x_1. \tag{3.25}$$

Here we used the following notations:

$$F = f\tilde{v}_{xx} + (f^2 - f_x)\tilde{v}_x, \qquad f = \frac{\varphi_x}{\varphi} = \frac{u_x(x, 0)}{u(x, 0)}, \qquad \tilde{\varepsilon}_1 = \frac{\varepsilon_1}{t}, \qquad \tilde{\varepsilon}_2 = \frac{\varepsilon_2}{t^3},$$

where ε_1 and ε_2 are the regularization parameters, t is the current instant of time, $v^*(x)$ is an arbitrary function, and $\tau_1 = t - \bar{t}$.

REMARKS ON CHOOSING THE PARAMETER τ_1

Remark 3.1. When simplifying equation (3.14) of the quasi-inversion method, the mean-value theorem was applied to the integrals

$$\int_0^t v(x, t)\,\mathrm{d}t, \qquad \int_0^t v_{x^{(i)}}(x, t)\,\mathrm{d}t, \quad i = 1, 2, 3.$$

It was assumed that a value of $\bar{t}$ is the same for all these equalities. Now we give the substantiation of this:

$$\int_0^t v(x, t)\,\mathrm{d}t = v(x, \bar{t})t, \qquad 0 < \bar{t} < t, \quad \bar{t} = \text{const},$$

$$\int_0^t v_{x^{(i)}}(x, t)\,\mathrm{d}t = \frac{\partial^i}{\partial x^i}\Big(\int_0^t v(x, t)\,\mathrm{d}t\Big) = \frac{\partial^i}{\partial x^i}(v(x, \bar{t})t) = v_{x^{(i)}}(x, \bar{t})t,$$

$$0 < t < \bar{t}, \qquad \bar{t} = \text{const}.$$

It is clear now that the value of $\bar{t}$ is the same both in the first and the second cases.

Remark 3.2 (choosing the value of τ_1). The following inequalities are obvious: $0 < \bar{t} < t < t_1$ and $0 < \tau_1 < t_1$ hold. Choosing the parameter τ_1, we use the mean-value theorem directly to determine $\bar{t}$ first, and then $\tau_1 = t - \bar{t}$. To find the value $\bar{t}$ we use either the formula of the mean-value theorem in the form $v(x,\bar{t})t = \int_0^t v(x,t)\,dt$ or the equality $[u_t(x,\bar{t})/u(x,\bar{t})]t = \ln[u(x,t)/\varphi(x)]$ recalling the way we introduce the function $v(x,t)$. A practical method for finding the value of the parameter τ_1 will be considered in the description of the numerical implementation of the developed algorithm.

3.5. FINDING THE COEFFICIENT

We shall determine the coefficient $k(x)$ analogously to the way we determined the coefficient $q(x)$ in Chapter 2. We consider several approaches to solution of this problem.

The first approach. We shall act as before when deriving equation (3.11). Integrate the original equation (3.7) with respect to t and write the result to solve for the derivative of the unknown coefficient:

$$k_x = \left(\int_0^t u_t\,dt - k\int_0^t u_{xx}\,dt\right)\Big/\int_0^t u_x\,dt.$$

Differentiate the obtained equality with respect to t observing that $k_{xt} = 0$:

$$(u_t - ku_{xx})\int_0^t u_x\,dt - \left(\int_0^t u_t\,dt - k\int_0^t u_{xx}\,dt\right)u_x = 0,$$

$$k = \frac{u_t\displaystyle\int_0^t u_x\,dt - u_x(u-\varphi)}{u_{xx}\displaystyle\int_0^t u_x\,dt - u_x\int_0^t u_{xx}\,dt} = \frac{\dfrac{u_t}{u}\displaystyle\int_0^t u_x\,dt - \dfrac{u_x}{u}(u-\varphi)}{\dfrac{u_{xx}}{u}\displaystyle\int_0^t u_x\,dt - \dfrac{u_x}{u}\int_0^t u_{xx}\,dt}.$$

As before, we introduce the function $v(x,t) = u_t/u$. With this way of introducing $v(x,t)$, we get

$$\frac{u_x}{u} = \int_0^t v_x\,dt + \frac{\varphi_x}{\varphi} = J_0, \qquad u = \varphi\exp\left(\int_0^t v\,dt\right) = \varphi(1-E),$$

$$\frac{u_{xx}}{u} = \left(\int_0^t v_x \, dt\right)^2 + 2\frac{\varphi_x}{\varphi}\int_0^t v_x \, dt + \int_0^t v_{xx} \, dt + \frac{\varphi_{xx}}{\varphi} = J_2,$$

$$\int_0^t u_x \, dt = \int_0^t \frac{u_x}{u} u \, dt = \int_0^t J_0 \varphi(1 - E) \, dt = \varphi J_1,$$

$$\int_0^t u_{xx} \, dt = \int_0^t \frac{u_{xx}}{u} u \, dt = \int_0^t J_2 \varphi(1 - E) \, dt = \varphi J_3, \tag{3.26}$$

Suppose that the function $\tilde{v}(x,t)$ is found as a result of solution of the quasi-inversion problem and the substitution $v(x,t) = \tilde{v}(x,t)$ is made. Then we get the formula for calculating the coefficient $k(x)$, which involves the solution of quasi-inversion problem:

$$\begin{aligned} k(x) &= \frac{vJ_1 + EJ_0}{J_1J_2 - J_0J_3} = \frac{A}{C} \\ &= \frac{-\bar{v}\bar{v}_x t^2/2}{[f\bar{v}_{xx} + (f^2 - f_x)\bar{v}_x]t^2/2} = \frac{\bar{v}\bar{v}_x}{(f^2 - f_x)\bar{v}_x - f\bar{v}_{xx}}. \end{aligned}$$

Finally we have

$$k = -\frac{\bar{v}\bar{v}_x}{F}. \tag{3.27}$$

Consider in greater detail the problem of finding the derivative $k_x(x)$ of the sought coefficient. There are different methods of finding it which lead to different formulas.

The first method consists in mere differentiation:

$$k_x(x) = \left(\frac{A}{C}\right)'_x = \frac{A_xC - AC_x}{C^2}.$$

We continue the calculations using the approximate representations for the coefficients A, C, and their derivatives:

$$\begin{aligned} k_x(x) = \{&-(\bar{v}_x^2 + \bar{v}\bar{v}_{xx})(f\bar{v}_{xx} + (f^2 - f_x)\bar{v}_x) + \bar{v}\bar{v}_x[f\bar{v}_{xxx} + f^2\bar{v}_{xx} \\ &+ (2ff_x - f_{xx})\bar{v}_x]\}/\{f\bar{v}_{xx} + [(f^2 - f_x)\bar{v}_x]^2\}. \end{aligned} \tag{3.28}$$

The second method of finding the derivative $k_x(x)$ is analogous to the derivation of the formula for $k(x)$. We integrate equation (3.7) with respect to t and write it to solve for the coefficient:

$$k = \left(\int_0^t u_t \, dt - k_x \int_0^t u_x \, dt\right) \Big/ \int_0^t u_{xx} \, dt$$

Taking into account that $k_t = 0$ we differentiate the obtained result with respect to t

$$(u_t - k_x u_x)\int_0^t u_{xx}\,\mathrm{dt} - \left(\int_0^t u_t\,\mathrm{dt} - k_x\int_0^t u_x\,\mathrm{dt}\right)u_{xx} = 0,$$

$$k_x = \frac{u_t\int_0^t u_{xx}\,\mathrm{dt} - u_{xx}\int_0^t u_t\,\mathrm{dt}}{u_x\int_0^t u_{xx}\,\mathrm{dt} - u_{xx}\int_0^t u_x\,\mathrm{dt}} = \frac{\frac{u_t}{u}\int_0^t u_{xx}\,\mathrm{dt} - \frac{u_{xx}}{u}(u-\varphi)}{\frac{u_x}{u}\int_0^t u_{xx}\,\mathrm{dt} - \frac{u_{xx}}{u_x}\int_0^t u_x\,\mathrm{dt}}$$
$$= \frac{vJ_3 + EJ_2}{-(J_1J_2 - J_0J_3)} = -C^{-1}(EJ_2 + vJ_3).$$

Using the approximate representation of the expressions that make up the obtained formula, we have

$$k_x = -\frac{(2f\bar{v}_x + \bar{v}_{xx})\bar{v}}{f\bar{v}_{xx} + (f^2 - f_x)\bar{v}_x} = -\frac{(2f\bar{v}_x + \bar{v}_{xx})\bar{v}}{F}. \tag{3.29}$$

Note that analytical comparing of the formulas for calculating $k_x(x)$ cannot be carried out. For this reason, the question of agreement between the results of calculations using these formulas will be clarified in the description of the numerical experiment.

Theorem 3.1. *After the substitution of the coefficient given by the formula* (3.27) *and its derivative given by* (3.29) *into equation* (3.2) *it becomes an identity.*

Proof. Substitute the values of the coefficient k and its derivative k_x given by (3.27) and (3.29), respectively into equation (3.2):

$$k\frac{u_{xx}}{u} + k_x\frac{u_x}{u} = \frac{u_t}{u}, \qquad \frac{A}{C}J_2 - \frac{EJ_2 + vJ_3}{C}J_0 = v,$$

$$AJ_2 - EJ_0J_2 - J_0J_3v = vC,$$

$$(EJ_0 + vJ_1)J_2 - EJ_0J_2 - J_0J_3v = J_1J_2v - J_0J_3v,$$

$$J_1J_2v - J_0J_3v \equiv J_1J_2v - J_0J_3v.$$

The resulting identity proves the assertion of the theorem. The obtained values of the sought coefficient and its derivative, being substituted into the equation derived by integrating equation (3.2) with respect to t, turn it also into an identity:

$$k\int_0^t u_{xx}\,\mathrm{dt} + k_x\int_0^t u_x\,\mathrm{dt} = \int_0^t u_t\,\mathrm{dt}, \qquad \frac{A}{C}\varphi J_3 - \frac{EJ_2+vJ_3}{C}\varphi J_1 = -\varphi E,$$

$$AJ_3 - EJ_1J_2 - vJ_1J_3 = -EC, \qquad (EJ_0+vJ_1)J_3 - EJ_1J_2 - vJ_1J_3 = -EC,$$

$$E(J_0J_3 - J_1J_2) = -EC, \qquad -EC \equiv -EC.$$

The theorem is proved. □

Remark. The form of equation (3.6) does not depend on the order in which the coefficient k and its derivative k_x are eliminated from equation (3.2). This assertion can be proved by a simple calculation.

Theorem 3.2. *The value of the derivative $k_x(x)$ of the unknown coefficient given by the formula (3.28) together with the coefficient $k(x)$ satisfy the original equation (3.2) and the equations derived from it.*

Proof. Transform the formula (3.28)

$$k(x) = \frac{A}{C} = \frac{EJ_0+vJ_1}{J_1J_2-J_0J_3} = \frac{\left(1-\frac{u}{\varphi}\right)\frac{u_x}{u} + \frac{u_t}{u}\frac{1}{\varphi}\int_0^t u_x\,\mathrm{dt}}{\frac{1}{\varphi}\int_0^t u_x\,\mathrm{dt}\,\frac{u_{xx}}{u} - \frac{u_x}{u}\frac{1}{\varphi}\int_0^t u_{xx}\,\mathrm{dt}}$$

$$= \frac{(\varphi-u)u_x + u_t\int_0^t u_x\,\mathrm{dt}}{u_{xx}\int_0^t u_x\,\mathrm{dt} - u_x\int_0^t u_{xx}\,\mathrm{dt}} = \frac{u_t\int_0^t u_x\,\mathrm{dt} - u_x\int_0^t u_t\,\mathrm{dt}}{u_{xx}\int_0^t u_x\,\mathrm{dt} - u_x\int_0^t u_{xx}\,\mathrm{dt}} = \frac{\Delta}{\Delta_1}$$

where

$$\Delta = \begin{vmatrix} \int_0^t u_x\,\mathrm{dt} & \int_0^t u_{xx}\,\mathrm{dt} \\ u_x & u_{xx} \end{vmatrix}, \qquad \Delta_1 = \begin{vmatrix} \int_0^t u_x\,\mathrm{dt} & \int_0^t u_t\,\mathrm{dt} \\ u_x & u_t \end{vmatrix}.$$

To calculate k_x we compose the determinants

$$\Delta_x = \begin{vmatrix} \int_0^t u_x\,\mathrm{dt} & \int_0^t u_{xxx}\,\mathrm{dt} \\ u_x & u_{xxx} \end{vmatrix},$$

$$\Delta_{1x} = \begin{vmatrix} \int_0^t u_x\,\mathrm{dt} & \int_0^t u_{xt}\,\mathrm{dt} \\ u_x & u_{xt} \end{vmatrix} + \begin{vmatrix} \int_0^t u_{xx}\,\mathrm{dt} & \int_0^t u_t\,\mathrm{dt} \\ u_{xx} & u_t \end{vmatrix}.$$

Then we get

$$k_x = \frac{\Delta_{1x}\Delta - \Delta_1\Delta_x}{\Delta^2} = \begin{vmatrix} \frac{1}{\Delta}\left(\begin{vmatrix} \int_0^t u_x\,\mathrm{dt} & \int_0^t u_{xt}\,\mathrm{dt} \\ u_x & u_{xt} \end{vmatrix} + \begin{vmatrix} \int_0^t u_{xx}\,\mathrm{dt} & \int_0^t u_t\,\mathrm{dt} \\ u_{xx} & u_t \end{vmatrix} \right) & \frac{\Delta}{\Delta_1} \\ \frac{1}{\Delta} \begin{vmatrix} \int_0^t u_x\,\mathrm{dt} & \int_0^t u_{xxx}\,\mathrm{dt} \\ u_x & u_{xxx} \end{vmatrix} & 1 \end{vmatrix}. \tag{3.30}$$

To analyze the formula (3.30), we consider the equation obtained by integrating of the original equation with respect to t and the equation obtained by differentiation of the original equation with respect to x

$$k_x u_x + k u_{xx} = u_t, \tag{3.31}$$

$$k_x \int_0^t u_x\,\mathrm{dt} + k \int_0^t u_{xx}\,\mathrm{dt} = \int_0^t u_t\,\mathrm{dt}, \tag{3.32}$$

$$2k_x u_{xx} + k u_{xxx} + k_{xx} u_x = u_{xt}, \tag{3.33}$$

$$2k_x \int_0^t u_{xx}\,\mathrm{dt} + k \int_0^t u_{xxx}\,\mathrm{dt} + k_{xx} \int_0^t u_x\,\mathrm{dt} = \int_0^t u_t\,\mathrm{dt}. \tag{3.34}$$

The coefficients of these equations include only the function $k(x)$ and its derivatives. We shall use this fact as follows. Consider equations (3.33) and (3.34) together and eliminate the coefficient k_{xx} from them:

$$2k_x \begin{vmatrix} \int_0^t u_{xx}\,\mathrm{dt} & \int_0^t u_x\,\mathrm{dt} \\ u_{xx} & u_x \end{vmatrix} + k \begin{vmatrix} \int_0^t u_{xxx}\,\mathrm{dt} & \int_0^t u_x\,\mathrm{dt} \\ u_{xxx} & u_x \end{vmatrix} = \begin{vmatrix} \int_0^t u_{xt}\,\mathrm{dt} & \int_0^t u_x\,\mathrm{dt} \\ u_{xt} & u_x \end{vmatrix}. \tag{3.35}$$

Transform equality (3.30). Multiply its both parts by Δ

$$k_x \begin{vmatrix} \int_0^t u_x\,\mathrm{dt} & \int_0^t u_{xx}\,\mathrm{dt} \\ u_x & u_{xx} \end{vmatrix} = \begin{vmatrix} \begin{vmatrix} \int_0^t u_x\,\mathrm{dt} & \int_0^t u_{xt}\,\mathrm{dt} \\ u_x & u_{xt} \end{vmatrix} + \begin{vmatrix} \int_0^t u_{xx}\,\mathrm{dt} & \int_0^t u_t\,\mathrm{dt} \\ u_{xx} & u_t \end{vmatrix} & k \\ \begin{vmatrix} \int_0^t u_x\,\mathrm{dt} & \int_0^t u_{xxx}\,\mathrm{dt} \\ u_x & u_{xxx} \end{vmatrix} & 1 \end{vmatrix}$$

Then we obtain

$$k_x \begin{vmatrix} \int_0^t u_x\,\mathrm{dt} & \int_0^t u_{xx}\,\mathrm{dt} \\ u_x & u_{xx} \end{vmatrix} = \begin{vmatrix} \int_0^t u_x\,\mathrm{dt} & \int_0^t u_{xt}\,\mathrm{dt} \\ u_x & u_{xt} \end{vmatrix} - \begin{vmatrix} \int_0^t u_t\,\mathrm{dt} & \int_0^t u_{xx}\,\mathrm{dt} \\ u_t & u_{xx} \end{vmatrix} - k \begin{vmatrix} \int_0^t u_x\,\mathrm{dt} & \int_0^t u_{xxx}\,\mathrm{dt} \\ u_x & u_{xxx} \end{vmatrix}. \tag{3.36}$$

Consider equalities (3.31), (3.32) together. Eliminate k from them, we get

$$k_x = \frac{1}{\Delta}\begin{vmatrix} \int_0^t u_t\,dt & \int_0^t u_{xx}\,dt \\ u_t & u_{xx} \end{vmatrix},$$

which implies

$$\begin{vmatrix} \int_0^t u_t\,dt & \int_0^t u_{xx}\,dt \\ u_t & u_{xx} \end{vmatrix} = k_x \begin{vmatrix} \int_0^t u_x\,dt & \int_0^t u_{xx}\,dt \\ u_x & u_{xx} \end{vmatrix}. \tag{3.37}$$

Substituting equality (3.37) into (3.36), we then bring it to the form (3.35). Theorem 2 is proved. □

The obtained coefficient $k(x)$ should satisfy the usual *condition of uniform parabolicity* (see Fridman, 1968). Now let us formulate this condition.

We rewrite equation (3.2) in the form

$$Lu = k(x)u_{xx} + k_x u_x - u_t = 0, \tag{3.38}$$

where the coefficient k is defined in the cylinder

$$\Omega \equiv \bar{D} \times [0, t_1] \equiv \{(x,t) : x \in \bar{D},\ 0 \le t \le t_1\}.$$

Here $\bar{D}$ is the closure of the bounded domain $D \subset \mathbb{R}^1$. The operator L is called parabolic at the point (x,t) if any real real vector ξ (in our case it is one-dimensional), $\xi = (\xi_1) \neq 0$ satisfies the condition $k(x)\xi_1^2 > 0$. If the operator L is parabolic at any points of Ω, then it is said to be parabolic in Ω.

If there exist the positive constants λ, γ such that any real vector ξ satisfies the condition

$$\lambda|\xi|^2 \le k(x)\xi_1^2 \le \gamma|\xi|^2 \tag{3.39}$$

for all $(x,t) \in \Omega$, then the operator L is uniformly parabolic in Ω. Obviously, condition (3.39) in the one-dimensional case has the form $\lambda \le k(x) \le \gamma$.

We now turn to analyzing the conditions under which neither the numerator nor denominator in the formula (3.27) vanish:

$$k(x) = \frac{\begin{vmatrix} \int_0^t u_x\,dt & \int_0^t u_t\,dt \\ u_x & u_t \end{vmatrix}}{\begin{vmatrix} \int_0^t u_x\,dt & \int_0^t u_{xx}\,dt \\ u_x & u_{xx} \end{vmatrix}} = \frac{\left(\dfrac{\int_0^t u_t\,dt}{\int_0^t u_x\,dt}\right)'_t \left(\displaystyle\int_0^t u_x\,dt\right)^2}{\left(\dfrac{\int_0^t u_{xx}\,dt}{\int_0^t u_x\,dt}\right)'_t \left(\displaystyle\int_0^t u_x\,dt\right)^2} = \frac{\left(\dfrac{\int_0^t u_t\,dt}{\int_0^t u_x\,dt}\right)'_t}{\left(\dfrac{\int_0^t u_{xx}\,dt}{\int_0^t u_x\,dt}\right)'_t}. \tag{3.40}$$

Let $\int_0^t u_x\,\mathrm{dt} \neq 0$, $\int_0^t u_{xx}\,\mathrm{dt} \neq 0$, and the condition $C \neq 0$ holds. We transform it as follows:

$$\begin{aligned}
C &= \begin{vmatrix} \int_0^t u_x\,\mathrm{dt} & \int_0^t u_{xx}\,\mathrm{dt} \\ u_x & u_{xx} \end{vmatrix} = u_{xx}\int_0^t u_x\,\mathrm{dt} - u_x\int_0^t u_{xx}\,\mathrm{dt} \\
&= \int_0^t u_x\,\mathrm{dt}\int_0^t u_{xx}\,\mathrm{dt}\left(\frac{u_{xx}}{\int_0^t u_{xx}\,\mathrm{dt}} - \frac{u_x}{\int_0^t u_x\,\mathrm{dt}}\right) \\
&= \int_0^t u_x\,\mathrm{dt}\int_0^t u_{xx}\,\mathrm{dt}\left[\left(\ln\left|\int_0^t u_{xx}\,\mathrm{dt}\right|\right)_t' - \left(\ln\left|\int_0^t u_x\,\mathrm{dt}\right|\right)_t'\right] \\
&= \int_0^t u_x\,\mathrm{dt}\int_0^t u_{xx}\,\mathrm{dt}\left(\ln\left|\frac{\int_0^t u_{xx}\,\mathrm{dt}}{\int_0^t u_x\,\mathrm{dt}}\right|\right)_t' \neq 0.
\end{aligned}$$

Hence $\left(\ln\left|\int_0^t u_{xx}\,\mathrm{dt}/\int_0^t u_x\,\mathrm{dt}\right|\right)_t' \neq 0$, which implies $\int_0^t u_{xx}\,\mathrm{dt}/\int_0^t u_x\,\mathrm{dt} = F(x,t)$, whereas $\int_0^t u_{xx}\,\mathrm{dt} \neq \int_0^t u_x\,\mathrm{dt}$ (this condition holds in the general case).

Let $\int_0^t u_x\,\mathrm{dt} \neq 0$, $\int_0^t u_t\,\mathrm{dt} \neq 0$. Then the condition $A \neq 0$ leads to the following result:

$$A = \begin{vmatrix} \int_0^t u_x\,\mathrm{dt} & \int_0^t u_t\,\mathrm{dt} \\ u_x & u_t \end{vmatrix} = \int_0^t u_x\,\mathrm{dt}\int_0^t u_t\,\mathrm{dt}\left(\ln\left|\frac{\int_0^t u_t\,\mathrm{dt}}{\int_0^t u_x\,\mathrm{dt}}\right|\right)_t' \neq 0.$$

Hence $\int_0^t u_t\,\mathrm{dt}/\int_0^t u_x\,\mathrm{dt} = \Phi(x,t)$ if $\int_0^t u_x\,\mathrm{dt} \neq \int_0^t u_t\,\mathrm{dt}$.

Theorem 3.3. *If $C = 0$ and $A = 0$, then equation (3.2) or (3.7) changes type in the solution domain. This follows not only from the form of equation (3.7), which is trivial, but also from the formulas obtained for the coefficients C and A.*

Proof. Suppose that $C = 0$ at some point $(x,t)$of the solution domain. With calculations similar to those carried out above, we conclude that, in this case, the following condition holds:

$$\left(\frac{\int_0^t u_{xx}\,\mathrm{dt}}{\int_0^t u_x\,\mathrm{dt}}\right)_t' = 0, \qquad \frac{\int_0^t u_{xx}\,\mathrm{dt}}{\int_0^t u_x\,\mathrm{dt}} = \psi(x), \qquad \int_0^t u_{xx}\,\mathrm{dt} = \psi(x)\int_0^t u_x\,\mathrm{dt}.$$

We transform this result differentiating it with respect to t

$$u_{xx} = \psi(x)u_x. \tag{3.41}$$

Substituting the dependence condition (3.41) into equation (3.2), we conclude that it changes type at the given point to become an equation of the first order

$$(k\psi(x) + k_x)u_x = u_t.$$

Suppose now that $A = 0$ at some point (x, t). In this case we obtain the following condition analogous to the previous one:

$$\left(\frac{\int_0^t u_t\,\mathrm{dt}}{\int_0^t u_x\,\mathrm{dt}}\right)'_t = 0, \qquad \frac{\int_0^t u_t\,\mathrm{dt}}{\int_0^t u_x\,\mathrm{dt}} = \zeta(x), \qquad \int_0^t u_t\,\mathrm{dt} = \zeta(x)\int_0^t u_x\,\mathrm{dt}.$$

Differentiate the result with respect to t

$$u_t = \zeta(x)u_x. \tag{3.42}$$

Substituting (3.42) into equation (3.2), we have

$$ku_{xx} + (k_x - \zeta(x))u_x = 0.$$

The equation type at the considered point (x, t) becomes elliptic. This proves the theorem. The points at which either $A = 0$ or $C = 0$ or both of these conditions hold together are degeneracy points of the original equation. □

Theorem 3.4. *For the values of the coefficient $k(x)$ given by the formula (3.27) to be positive, it is necessary that the condition $u_{xx}u_t > 0$ hold, i. e., the derivatives u_{xx} and u_t of the solution should have the same sign in the solution domain of the initial-boundary value problem for equation (3.1).*

Proof. We transform the formula (3.27) assuming that

$$\int_0^t u_x\,\mathrm{dt} \neq 0, \qquad \int_0^t u_{xx}\,\mathrm{dt} \neq 0, \qquad \int_0^t u_t\,\mathrm{dt} \neq 0.$$

We have

$$k = \frac{u_t\int_0^t u_x\,\mathrm{dt} - u_x\int_0^t u_t\,\mathrm{dt}}{u_{xx}\int_0^t u_x\,\mathrm{dt} - u_x\int_0^t u_{xx}\,\mathrm{dt}} = \frac{\int_0^t u_x\,\mathrm{dt}\int_0^t u_t\,\mathrm{dt}\left[\frac{u_t}{\int_0^t u_t\,\mathrm{dt}} - \frac{u_x}{\int_0^t u_x\,\mathrm{dt}}\right]}{\int_0^t u_{xx}\,\mathrm{dt}\int_0^t u_x\,\mathrm{dt}\left[\frac{u_{xx}}{\int_0^t u_{xx}\,\mathrm{dt}} - \frac{u_x}{\int_0^t u_x\,\mathrm{dt}}\right]}$$

$$= \frac{\int_0^t u_x\,\mathrm{dt}\left(\ln\frac{\int_0^t u_t\,\mathrm{dt}}{\int_0^t u_x\,\mathrm{dt}}\right)'_t}{\int_0^t u_{xx}\,\mathrm{dt}\left(\ln\frac{\int_0^t u_{xx}\,\mathrm{dt}}{\int_0^t u_x\,\mathrm{dt}}\right)'_t} = \frac{\int_0^t u_t\,\mathrm{dt}(\ln\Phi(x,t))'_t}{\int_0^t u_{xx}\,\mathrm{dt}(\ln F(x,t))'_t}.$$

Consider the functions $\ln \Phi(x,t)$ and $\ln F(x,t)$, which are monotonically increasing for any fixed x. Consequently,

$$(\ln \Phi(x,t))'_t > 0, \qquad (\ln F(x,t))'_t > 0.$$

The condition $k(x) > 0$ leads to inequality

$$\int_0^t u_{xx}\,\mathrm{d}t \int_0^t u_t\,\mathrm{d}t > 0. \tag{3.43}$$

Using the properties of defined integrals, we rewrite (3.43) in the form

$$u_{xx}u_t > 0. \tag{3.44}$$

The theorem is proved. □

The assertion of Theorem 3.4 is in agreement with the usual conditions for the special case $k(x) = \text{const}$. Inequality (3.44) makes it possible to rewrite equation (3.46) in the usual form (see Tikhonov and Samarskii, 1966; Krikunov, 1970)

$$a^2 u_{xx} = u_t,$$

which implies the derivatives u_{xx} and u_t should have the same sign in the solution domain of the initial-boundary value problem for this equation.

The second approach. When determining the coefficient $k(x)$ we shall not integrate equation (3.2) with respect to time as we did before. Write equation (3.2) to solve for the derivative $k_x = (u_t - ku_{xx})/u_x$, differentiate the obtained equality with respect to t and observe that $k_{xt} = 0$. Write the result to solve for k:

$$k = \frac{u_{xt}u_t - u_x u_{tt}}{u_{xt}u_{xx} - u_x u_{xxt}}. \tag{3.45}$$

In the formula (3.45), introduce the function $v = \tilde{v}(x,t)$ found as a result of solution of the corresponding quasi-inversion problem. Using the transfer formulas (3.26) and the following relations:

$$\frac{u_{tt}}{u} = v^2 + v_t, \qquad \frac{u_{xt}}{u} = J_0 v + v_x, \qquad \frac{u_{xxt}}{u} = J_2 v + 2J_0 v_x + v_{xx}$$

we obtain

$$k = \frac{\tilde{v}_x \tilde{v} - J_0 \tilde{v}_t}{(J_2 - 2J_0^2)\tilde{v}_x - J_0 \tilde{v}_{xx}}. \tag{3.46}$$

Now eliminate the coefficient $k(x)$ from equation (3.2) and write the result to solve for its derivative $k_x(x)$:

$$k_x = \frac{u_{xxt}u_t - u_{xx}u_{tt}}{u_{xxt}u_x - u_{xt}u_{xx}}. \tag{3.47}$$

After introducing the function $\tilde{v}(x,t)$, the formula (3.47) takes the form

$$k_x = \frac{\tilde{v}_{xx}\tilde{v} + 2J_0\tilde{v}_x\tilde{v} - J_2\tilde{v}_t}{(J_2 - 2J_0^2)\tilde{v}_x - J_0\tilde{v}_{xx}}. \tag{3.48}$$

Theorem 3.5. *After the substitution of the coefficient $k(x)$ and its derivative $k_x(x)$ given by the formulas (3.27) and (3.28), respectively, the following equation obtained by integrating the original equation (3.2) with respect to t becomes an identity:*

$$k\int_0^t u_{xx}\,\mathrm{d}t + k_x\int_0^t u_x\,\mathrm{d}t = \int_0^t u_t\,\mathrm{d}t. \tag{3.49}$$

Proof. In equation (3.49), introduce the function $v(x,t)$

$$kJ_3 + k_xJ_1 = -E. \tag{3.50}$$

Substitute the formulas (3.46), (3.48) into (3.50):

$$\frac{\tilde{v}_x\tilde{v} - J_0\tilde{v}_t}{(J_2 - 2J_0^2)\tilde{v}_x - J_0\tilde{v}_{xx}}J_3 - \frac{\tilde{v}_{xx}\tilde{v} + 2J_0\tilde{v}_x\tilde{v} - J_2\tilde{v}_t}{(J_2 - 2J_0^2)\tilde{v}_x - J_0\tilde{v}_{xx}} = -E,$$

$$\tilde{v}_x\tilde{v}J_3 - J_0\tilde{v}_tJ_3 - J_1\tilde{v}_{xx}\tilde{v} + 2J_0J_1\tilde{v}_x\tilde{v} - J_1J_2\tilde{v}_t = -E(J_2 - 2J_0^2)\tilde{v}_x + EJ_0\tilde{v}_{xx}.$$

Grouping the terms in the resulting equality, we obtain

$$(J_1J_2 - J_0J_3)\tilde{v}_t = (EJ_0 + \tilde{v}J_1)\tilde{v}_{xx} + [E(2J_0^2 - J_2) + \tilde{v}(2J_0J_1 - J_3)]\tilde{v}_x.$$

Rewrite this expression using the notation introduced above in the form

$$C\tilde{v}_t = A\tilde{v}_{xx} + B\tilde{v}_x. \tag{3.51}$$

Note that equation (3.51) is the original equation for finding the function $v(x,t)$. Since the solution of this equation is close to the solution of the corresponding quasi-inversion problem, we can conclude that (3.51) is an identity. This proves the theorem. □

Thus, the coefficient $k(x)$ and its derivative determined from equation (3.2) will also satisfy this equation after it is integrated with respect to t.

Transform the formula (3.48) using the simplification made in Section 3.3. We obtain

$$k = \frac{vv_x - fv_t}{-F}, \qquad k_x = \frac{(2fv_x + v_{xx})v - (f_x + f^2)v_t}{F}. \tag{3.52}$$

This result differs from the one obtained before (see (3.27)). However, the value of the derivative v_t is small since, when using the quasi-inversion method, the interval of integration of equation (3.2) with respect to t is small. Therefore, we can say that these formulas approximately coincide.

Theorem 3.6. *The formulas* (3.27) *and* (3.52) *give the same result up to the infinitesimal terms of order* $O(t^3)$.

Proof. The formula (3.52) differs from the formula (3.27) by the summand $-fv_t$ in the numerator. Write equation (3.7) with the derivative v_t on the left-hand side and, using the simplification in Section 3.3, find

$$v_t = \frac{A}{C} v_{xx} + \frac{B}{C} v_x = \frac{-vv_x}{F} v_{xx} + \frac{vv_{xx}}{F} v_x = 0.$$

Simplifications carried out in Section 3.4 which were used here are obtained with accuracy of $O(t^3)$. This is the accuracy of coincidence of the formulas (3.27) and (3.52). □

3.6. DIFFERENCE QUASI-INVERSION PROBLEM

Write the quasi-inversion problem in the finite-difference form. Introduce the mesh with lengths $h = (x_1 - x_0)/J$, $\tau = t_1/N$ and consider the mesh function $v_i^n = \tilde{v}(ih, n\tau)$. As usual, here J and N are the numbers of mesh cells with respect to the variables x and t, respectively. Using finite differences of the second order of accuracy, we have

$$\left(\frac{\partial \tilde{v}}{\partial t}\right)_i^n = \tilde{v}_{\cot t} + O(\tau^2) = \frac{\tilde{v}_i^{n+1} - \tilde{v}_i^{n-1}}{2\tau} + O(\tau^2),$$

$$\left(\frac{\partial^2 \tilde{v}}{\partial t^2}\right)_i^n = \tilde{v}_{\bar{t}t} + O(\tau^2) = \frac{\tilde{v}_i^{n+1} - 2\tilde{v}_i^n + \tilde{v}_i^{n-1}}{\tau^2} + O(\tau^2),$$

$$\left(\frac{\partial \tilde{v}}{\partial x}\right)_i^n = \tilde{v}_{\cot x} + O(h^2) = \frac{\tilde{v}_{i+1}^n - \tilde{v}_{i-1}^n}{2\tau} + O(h^2),$$

$$\left(\frac{\partial^2 \tilde{v}}{\partial x^2}\right)_i^n = \tilde{v}_{\bar{x}x} + O(h^2) = \frac{\tilde{v}_{i+1}^n - 2\tilde{v}_i^n + \tilde{v}_{i-1}^n}{h^2} + O(h^2),$$

$$\left(\frac{\partial^3 \tilde{v}}{\partial x^3}\right)_i^n = \frac{\tilde{v}_{i+2}^n - 2\tilde{v}_{i+1}^n + 2\tilde{v}_{i-1}^n - \tilde{v}_{i-2}^n}{2h^3} + O(h^2),$$

$$\left(\frac{\partial^4 \tilde{v}}{\partial x^4}\right)_i^n = \frac{\tilde{v}_{i+2}^n - 4\tilde{v}_{i+1}^n + 6\tilde{v}_i^n - 4\tilde{v}_{i-1}^n + \tilde{v}_{i-2}^n}{h^4} + O(h^2),$$

$$\left(\frac{\partial^2 \tilde{v}}{\partial x \partial t}\right)_i^n = \frac{\tilde{v}_{i+1}^{n+1} - \tilde{v}_{i-1}^{n+1} - \tilde{v}_{i+1}^{n-1} + \tilde{v}_{i-1}^{n-1}}{4h\tau} + O(h^2 + \tau^2),$$

$$\left(\frac{\partial^5 \tilde{v}}{\partial t \partial x^4}\right)_i^n = \frac{1}{2h^4\tau}\Big[(\tilde{v}_{i+2}^{n+1} - 4\tilde{v}_{i+1}^{n+1} + 6\tilde{v}_i^{n+1} - 4\tilde{v}_{i-1}^{n+1} + \tilde{v}_{i-2}^{n+1}) \\ - (\tilde{v}_{i+2}^{n-1} - 4\tilde{v}_{i+1}^{n-1} + 6\tilde{v}_i^{n-1} - 4\tilde{v}_{i-1}^{n-1} + \tilde{v}_{i-2}^{n-1})\Big] + O(h^2 + \tau^2).$$

Rewrite the equation of modified quasi-inversion problem (3.25) collecting the second summand and the fifth, i.e., the last summand on the left-hand side:

$$-(\tilde{v}^2\tilde{v}_x\tilde{v}_{xt}\tau_1 t)\tilde{v}_{xxxx} + [\tilde{v}^2\tilde{v}_{xt}(\tilde{v}_{xxxxt}\tau_1 + 6F\tilde{v}_x t)]\tilde{v}_x \\ + (F^2 + 4\tilde{\varepsilon}_1\tilde{\varepsilon}_2)t\tilde{v}_{tt} + (2F - \tilde{\varepsilon}_1)2F\tilde{v}_t = 0, \tag{3.53}$$

where $\tilde{\varepsilon}_1 = \varepsilon_1/t$ and $\tilde{\varepsilon}_2 = \varepsilon_2/t^3$. Denote

$$a_i^n = -(\tilde{v}^2\tilde{v}_x\tilde{v}_{xt}\tau_1 t)_i^n, \qquad b_i^n = (\tilde{v}^2\tilde{v}_{xt}(\tilde{v}_{xxxxt}\tau_1 + 6F\tilde{v}_x t))_i^n,$$

$$c_i^n = ((F^2 + 4\tilde{\varepsilon}_1\tilde{\varepsilon}_2)t)_i^n, \qquad d_i^n = ((2F - \tilde{\varepsilon}_1)2F)_i^n,$$

where a_i^n, b_i^n, c_i^n, d_i^n are difference coefficients of equation (3.53). Write equation (3.53) in the difference form:

$$a_i^n \frac{\tilde{v}_{i+2}^n - 4\tilde{v}_{i+1}^n + 6\tilde{v}_i^n - 4\tilde{v}_{i-1}^n + \tilde{v}_{i-2}^n}{h^4} + b_i^n \frac{\tilde{v}_{i+1}^n - \tilde{v}_{i-1}^n}{2h} \\ + c_i^n \frac{\tilde{v}_i^{n+1} - 2\tilde{v}_i^n + \tilde{v}_i^{n-1}}{\tau^2} + d_i^n \frac{\tilde{v}_i^{n+1} - \tilde{v}_i^{n-1}}{2\tau} = 0. \tag{3.54}$$

The difference equation (3.54) (without regard for the difference expression of its coefficients) is based on the following pattern:

$$\begin{array}{ccccc} & & {}^*(i, n+1) & & \\ {}^*(i-2, n) & {}^*(i-1, n) & {}^*(i, n) & {}^*(i+1, n) & {}^*(i+2, n) \\ & & {}^*(i, n-1). & & \end{array}$$

In equation (3.54), we group the values of the sought function at the mesh points:

$$\frac{a_i^n}{h^4}\tilde{v}_{i+2}^n + \Big(-4\frac{a_i^n}{h^4} + \frac{b_i^n}{2h}\Big)\tilde{v}_{i+1}^n + \Big(6\frac{a_i^n}{h^4} - 2\frac{c_i^n}{\tau^2}\Big)\tilde{v}_i^n + \Big(-4\frac{a_i^n}{h^4} + \frac{b_i^n}{2h}\Big)\tilde{v}_{i-1}^n \\ + \frac{a_i^n}{h^4}\tilde{v}_{i-2}^n + \Big(\frac{c_i^n}{\tau^2} + \frac{d_i^n}{2\tau}\Big)\tilde{v}_i^{n+1} + \Big(\frac{c_i^n}{\tau^2} - \frac{d_i^n}{2\tau}\Big)\tilde{v}_i^{n-1} = 0. \tag{3.55}$$

Introducing obvious notations, we rewrite equation (3.55) in the form

$$A_i^n\tilde{v}_{i+2}^n + B_i^n\tilde{v}_{i+1}^n + C_i^n\tilde{v}_i^n + D_i^n\tilde{v}_{i-1}^n + M_i^n\tilde{v}_{i-2}^n + N_i^n\tilde{v}_i^{n+1} + K_i^n\tilde{v}_i^{n-1} = 0. \tag{3.56}$$

The system of difference equations (3.56) is written for $1 \le j \le J_1$ and $1 \le i \le N-1$. It is solvable due to the boundary conditions at $x = x_0$, $x = x_1$, $t = 0$, and $t = t_1$.

The difference form of the boundary conditions (3.16) and (3.17) of the problem (3.25) is

$$\tilde{v}_0^n = \mu_0^n, \qquad \tilde{v}_{-1}^n = \tilde{v}_1^n - 2h\eta_0^n, \tag{3.57}$$

$$\tilde{v}_J^n = \mu_1^n, \qquad \tilde{v}_{J+1}^n = \tilde{v}_{J-1}^n + 2h\eta_1^n. \tag{3.58}$$

Before writing the difference form of the condition (3.21), following Lattes and Lions (1967) we transform it into the form

$$\tilde{v}(x,0) - \varepsilon_2 \tilde{v}_t(x,0) = 0. \tag{3.59}$$

Thus, the arbitrary function $v^*(x)$ is chosen to be zero. In Lattes and Lions (1967) it is proved that the solution of quasi-inversion problem approximates the sought function whatever the function $v^*(x)$ may be, of course,outside the neighbourhood of the point $t = 0$.

Theorem 3.7. *Regardless of the form of the boundary condition for $t = 0$ (either (3.21) or (3.59)), it means that $t = 0$ the function $\tilde{v}(x,0)$ is specified arbitrarily.*

Proof. Condition (3.21) can be viewed as a linear ordinary differential equation

$$\varepsilon_2 \tilde{v}_t(x,0) - \tilde{v}(x,0) = -v^*$$

for the function $\tilde{v}(x,t)$, where the variable x serves as a parameter. Solving this equation, we find

$$\tilde{v}(x,t) = v^*(x) + C(x) \exp\left(\frac{t}{\varepsilon_2}\right),$$

where $C(x)$ is the integration constant depending on the "parameter" x. Hence, for $t = 0$ we get

$$\tilde{v}(x,t) = v^*(x) + C(x) \tag{3.60}$$

□

Consider the case where the boundary condition for $t = 0$ is given in the form (3.60). Then $\tilde{v}(x,0) = C(x)$, where the function $C(x)$ should satisfy the boundary conditions for $x = x_0$, $x = x_1$:

$$C(x_0) = \mu_0(0), \qquad \frac{\mathrm{d}C}{\mathrm{d}x}(x_0) = \eta_0(0)$$

$$C(x_1) = \mu_1(0), \qquad \frac{dC}{dx}(x_1) = \eta_1(0)$$

Analyze the procedure to specify the function $C(x)$. We operate on the assumption that it is an arbitrary function satisfying the boundary conditions. For $C(x)$ in the neighbourhood of the point x_0 we take its Taylor series expansion with accuracy of $O(h^2)$:

$$C(x) = \mu_0(0) + \eta_0(0)(x - x_0) + O((x - x_0)^2), \quad x = x_1 - h.$$

We assume the graph of the function $C(x)$ on the interval $(x_0 + h, x_1 - h)$ to be straight line passing through the points $(x_0 + h, \mu_0(0) + \eta_0(0)h)$ and $(x_1 - h, \mu_1(0) - \eta_1(0)h)$:

$$C(x) = \mu_0(0) + \eta_0(0)h + \frac{(\mu_1(0) - \mu_0(0)) - (\eta_1(0) + \eta_0(0))h}{x_1 - x_0 - 2h}(x - x_0 - h).$$

Consider the difference form of the boundary condition for $t = t_1$. Recall the way this problem was solved in Lattes and Lions (1970). To approximate the derivative with respect to t, this boundary condition involves the finite-difference analogue of the first order of accuracy

$$\left(\frac{\partial \tilde{v}}{\partial t}\right)_i^N = \frac{\tilde{v}_i^N - \tilde{v}_i^{N-1}}{\tau} + O(\tau) = \tilde{v}_{\bar{t}} + O(\tau) = \frac{\tilde{v} - \hat{\tilde{v}}}{\tau} + O(\tau).$$

The boundary condition is associated with various difference relations, depending on the way of approximation of other derivatives involved. Three variants of difference forms of these derivatives were studied. We shall act similarly, though it should be noted that using a difference analogue of the first order of accuracy for the derivative $\tilde{v}_t$ does not comply with the general procedure of approximation of the equation and the other boundary conditions, which may lead to greater errors in the solution. This can be avoided by using a nonuniform mesh with respect to t. The idea is as follows: for the uniform mesh with length τ we choose the last spacing τ_1 to satisfy the condition $\tau_1 = \tau^2$. In this case the order of approximation is different, but the accuracy is preserved.

Similarly to the procedure in Lattes and Lions (1967), we associate the non-linear boundary condition (3.23) with three different finite-difference expressions. They represent three ways of constructing difference analogues of the derivative $(\tilde{v}_x)_i^N$

$$\left(\frac{\partial \tilde{v}}{\partial x}\right)_i^N = \frac{\tilde{v}_{i+1}^N - \tilde{v}_{i-1}^N}{2h} + O(h^2),$$

$$\left(\frac{\partial \tilde{v}}{\partial x}\right)_i^N = \frac{1}{2}\left(\frac{\tilde{v}_{i+1}^N - \tilde{v}_{i-1}^N}{2h} + \frac{\tilde{v}_{i+1}^{N-1} - \tilde{v}_{i-1}^{N-1}}{2h}\right) + O(h^2),$$

$$\left(\frac{\partial \tilde{v}}{\partial x}\right)_i^N = \frac{1}{2}\left(\frac{\tilde{v}_{i+2}^N + \tilde{v}_{i+1}^N - \tilde{v}_{i-1}^N - \tilde{v}_{i-2}^N}{6h} + \frac{\tilde{v}_{i+2}^{N-1} + \tilde{v}_{i+1}^{N-1} - \tilde{v}_{i-1}^{N-1} - \tilde{v}_{i-2}^{N-1}}{6h}\right) + O(h^2)$$

and three ways of constructing difference analogues of the second derivative $(\tilde{v}_{xx})_i^N$

$$\left(\frac{\partial^2 \tilde{v}}{\partial x^2}\right)_i^N = \frac{\tilde{v}_{i+1}^N - 2\tilde{v}_i^N + \tilde{v}_{i-1}^N}{h^2} + O(h^2),$$

$$\left(\frac{\partial^2 \tilde{v}}{\partial x^2}\right)_i^N = \frac{1}{2}\left(\frac{\tilde{v}_{i+1}^N - 2\tilde{v}_i^N + \tilde{v}_{i-1}^N}{h^2} + \frac{\tilde{v}_{i+1}^{N-1} - 2\tilde{v}_i^{N-1} + \tilde{v}_{i-1}^{N-1}}{h^2}\right) + O(h^2),$$

$$\left(\frac{\partial^2 \tilde{v}}{\partial x^2}\right)_i^N = \frac{1}{2}\left(\frac{-\tilde{v}_{i+2}^N + 16\tilde{v}_{i+1}^N - 30\tilde{v}_i^N + 16\tilde{v}_{i-1}^N - \tilde{v}_{i-2}^N}{12h^2} + \frac{-\tilde{v}_{i+2}^{N-1} + 16\tilde{v}_{i+1}^{N-1} - 30\tilde{v}_i^{N-1} + \tilde{v}_{i-1}^{N-1} - \tilde{v}_{i-2}^{N-1}}{12h^2}\right) + O(h^2)$$

Using these difference analogues, we present three variants of difference forms of the boundary condition (3.23) and the corresponding patterns.

The first variant:

$$(2\varepsilon_1\varepsilon_2 + F_i^N t_1^2)\,\frac{\tilde{v}_i^N - \tilde{v}_i^{N-1}}{\tau} + (\tilde{v}_i^N)^2\,\frac{\tilde{v}_{i+1}^N - \tilde{v}_{i-1}^N}{8h^3}\,t_1^2 = 0, \tag{3.61}$$

where

$$F_i^N = f_i\,\frac{1}{h^2}(\tilde{v}_{i+1}^N - 2\tilde{v}_i^N + \tilde{v}_{i-1}^N) + \left(f_i^2 - \frac{f_{i+1} - f_{i-1}}{2h}\right)\frac{\tilde{v}_{i+1}^N - \tilde{v}_{i-1}^N}{2h},$$

$$f_i = \frac{\varphi_{i+1} - \varphi_{i-1}}{\varphi_i}\,\frac{1}{2h}.$$

and the following pattern is used:

$$\begin{array}{cccc} {}^*(i-1,N) & {}^*(i,N) & & {}^*(i+1,N) \\ & {}^*(i,N-1) & & \end{array}$$

The second variant

$$(2\varepsilon_1\varepsilon_2 + F_i^N t_1^2)\frac{\tilde v_i^N - \tilde v_i^{N-1}}{\tau} + (\tilde v_i^N)^2 \frac{1}{8}\Big(\frac{\tilde v_{i+1}^N - \tilde v_{i-1}^N}{2h} + \frac{\tilde v_{i+1}^{N-1} - \tilde v_{i-1}^{N-1}}{2h}\Big)^3 t_1^4 = 0, \tag{3.62}$$

where

$$F_i^N = \frac{f_i}{2}\Big(\frac{\tilde v_{i+1}^N - 2\tilde v_i^N + \tilde v_{i-1}^N}{h^2} + \frac{\tilde v_{i+1}^{N-1} - 2\tilde v_i^{N-1} + \tilde v_{i-1}^{N-1}}{h^2}\Big) + \frac{1}{2}\Big(f_i^2 - \frac{f_{i+1} - f_{i-1}}{2h}\Big)\Big(\frac{\tilde v_{i+1}^N - \tilde v_{i-1}^N}{2h} + \frac{\tilde v_{i+1}^{N-1} - \tilde v_{i-1}^{N-1}}{2h}\Big).$$

and the following pattern is used:

$$\begin{matrix} {}^*(i-1,N) & {}^*(i,N) & {}^*(i+1,N) \\ {}^*(i-1,N-1) & {}^*(i,N-1) & {}^*(i+1,N-1) \end{matrix}$$

The third variant

$$(2\varepsilon_1\varepsilon_2 + F_i^N t_1^2)\frac{\tilde v_i^N - \tilde v_i^{N-1}}{\tau} + \frac{(\tilde v_i^N)^2}{2}\Big(\frac{\tilde v_{i+2}^N - \tilde v_{i+1}^N - \tilde v_{i-1}^N - \tilde v_{i-2}^N}{6h} + \frac{\tilde v_{i+2}^{N-1} - \tilde v_{i+1}^{N-1} - \tilde v_{i-1}^{N-1} - \tilde v_{i-2}^{N-1}}{6h}\Big)^3 t_1^4 = 0, \tag{3.63}$$

where

$$F_i^N = \frac{f_i}{2}\Big(\frac{-\tilde v_{i+2}^N + 16\tilde v_{i+1}^N - 30\tilde v_i^N + 16\tilde v_{i-1}^N - \tilde v_{i-2}^N}{12h^2} + \frac{-\tilde v_{i+2}^{N-1} + 16\tilde v_{i+1}^{N-1} - 30\tilde v_i^{N-1} + 16\tilde v_{i-1}^{N-1} - \tilde v_{i-2}^{N-1}}{12h^2}\Big) + \Big(f_i^2 - \frac{f_{i+1} - f_{i-1}}{2h}\Big)\frac{1}{2} \times \Big(\frac{\tilde v_{i+2}^N + \tilde v_{i+1}^N - \tilde v_{i-1}^N - \tilde v_{i-2}^N}{6h} + \frac{\tilde v_{i+2}^{N-1} + \tilde v_{i+1}^{N-1} - \tilde v_{i-1}^{N-1} - \tilde v_{i-2}^{N-1}}{6h}\Big).$$

and the following pattern is used:

$$\begin{matrix} {}^*(i-2,N) & {}^*(i-1,N) & {}^*(i,N) & {}^*(i+1,N) & {}^*(i+2,N) \\ {}^*(i-2,N-1) & {}^*(i-1,N-1) & {}^*(i,N-1) & {}^*(i+1,N-1) & {}^*(i+2,N-1) \end{matrix}$$

3.7. NUMERICAL SOLUTION OF THE QUASI-INVERSION PROBLEM

Introduce the column vector of variables

$$\vec{V} = \begin{pmatrix} \tilde{v}_1^n \\ \tilde{v}_2^n \\ \cdot \\ \tilde{v}_{J-1}^n \end{pmatrix}.$$

The system of equation (3.56) together with the relations (3.57), (3.58), (3.60), and one of the relations (3.61)–(3.63) can be written in the matrix form

$$\mathbf{A}\vec{V} = \vec{F}$$

where the matrix **A** depends on the choice of the conditions on the boundary $t = t_1$. It is seven-diagonal for (3.61) and (3.62) or nine-diagonal for (3.63). The values of the unknown function entering the coefficients of the boundary condition (3.23) are assumed to be known and are specified in the usual way in the beginning of the iterative process.

We associate the difference problem of the quasi-inversion method with the system of three-point vector equations

$$\mathbf{A}\vec{V}_{n-1} - \mathbf{C}\vec{V}_n + \mathbf{B}\vec{V}_{n-1} = -\vec{F}_n, \qquad n = 1, \ldots, N-1. \tag{3.64}$$

As usual, to solve the system (86) we use the matrix sweep method (see Samarskii and Nikolaev 1978, pp. 107–112). Introduce the matrices

$$\mathbf{A} = \begin{pmatrix} K_1^n & 0 & \cdot & 0 \\ 0 & K_2^n & \cdot & 0 \\ \cdot & \cdot & \cdot & \cdot \\ 0 & 0 & \cdot & K_{J-1}^n \end{pmatrix}_{(J-1)\times(J-1)}, \qquad \mathbf{B} = \begin{pmatrix} N_1^n & 0 & \cdot & 0 \\ 0 & N_2^n & \cdot & 0 \\ \cdot & \cdot & \cdot & \cdot \\ 0 & 0 & \cdot & N_{J-1}^n \end{pmatrix}_{(J-1)\times(J-1)},$$

$$K_i^n = \frac{c_i^n}{\tau^2} - \frac{d_i^n}{2\tau}, \qquad N_i^n = \frac{c_i^n}{\tau^2} + \frac{d_i^n}{2\tau},$$

$$
\mathbf{C} = \begin{pmatrix}
-(C_1^n + M_1^n) & -B_1^n & -A_1^n & 0 & 0 & 0 & \cdot & \cdot & 0 & 0 & 0 & 0 & 0 \\
-D_2^n & -C_2^n & -B_2^n & -A_2^n & 0 & 0 & \cdot & \cdot & 0 & 0 & 0 & 0 & 0 \\
-M_3^n & -D_3^n & -C_3^n & -B_3^n & -A_3^n & 0 & \cdot & \cdot & 0 & 0 & 0 & 0 & 0 \\
0 & -M_4^n & -D_4^n & -C_4^n & -B_4^n & -A_4^n & \cdot & \cdot & \cdot & \cdot & \cdot & \cdot & \cdot \\
\cdot & \cdot & \cdot & \cdot & \cdot & \cdot & \cdot & \cdot & -M_{J-3}^n & -D_{J-3}^n & -C_{J-3}^n & -B_{J-3}^n & -A_{J-3}^n \\
0 & 0 & 0 & 0 & 0 & 0 & 0 & \cdot & 0 & -M_{J-2}^n & -D_{J-2}^n & -C_{J-2}^n & -B_{J-2}^n \\
0 & 0 & 0 & 0 & 0 & 0 & 0 & \cdot & 0 & 0 & -M_{J-1}^n & -D_{J-1}^n & -(A_{J-1}^n + C_{J-1}^n) \\
0 & 0 & 0 & 0 & 0 & 0 & 0 & & & & & &
\end{pmatrix}_{(J-1)\times(J-1)}
$$

$$
A_i^n = M_i^n = \frac{a_i^n}{h^4} = -\frac{(\tilde{v}^2 \tilde{v}_x \tilde{v}_{xt} \tau_1 t)_i^n}{h^4}, \qquad B_i^n = -\frac{4a_i^n}{h^4} + \frac{b_i^n}{2h},
$$

$$
C_i^n = \frac{6a_i^n}{h^4} - \frac{2c_i^n}{\tau^2}, \qquad D_i^n = -\frac{4a_i^n}{h^4} - \frac{b_i^n}{2h},
$$

$$
\vec{F}_n = \begin{pmatrix}
D_1^n \mu_0^n - M_1^n 2h\eta_0^n \\
M_2^n \mu_0^n \\
\left.\begin{matrix} 0 \\ \vdots \\ 0 \end{matrix}\right\} J-5 \\
A_{j-2}^n \mu_1^n \\
A_{J-1}^n 2h\eta_1^n + B_{J-1}^n \mu_1^n
\end{pmatrix}_{(J-1)} .
$$

Supply the system of equations (3.64) with boundary conditions. For $n = 0$ we have

$$
-\mathbf{C}_0 \vec{V}_0 + \mathbf{B}_0 \vec{V}_1 = -\vec{F}_0. \tag{3.65}
$$

Let the condition on the boundary (for $t = 0$) be

$$
\tilde{v}(x, 0) = v^*(x). \tag{3.66}
$$

where the function $v^*(x)$ is concordant with the boundary conditions and chosen as follows:

$$v^*(x) = \begin{cases} \mu_0(0), & x = x_0 \\ \mu_0(0) + \eta_0(0)h, & x = x_0 + h \\ \mu_0(0) + \eta_0(0)h \\ \quad + \dfrac{(\mu_1(0) - \mu_0(0)) - (\eta_1(0) - \eta_0(0))h}{x_1 - x_0 - 2h}(x_1 - x_0 - h), \\ & x_0 + h < x < x_1 - h \\ \mu_1(0) + \eta_1(0)h, & x = x_1 - h \\ \mu_1(0), & x = x_1. \end{cases}$$

We rewrite the boundary condition (3.66) in the form $\vec{V}_0 = \vec{F}_0$, where the matrix has the form

$$\vec{F}_0 = \begin{pmatrix} v_1^* \\ v_2^* \\ \cdot \\ v_{J-1}^* \end{pmatrix}.$$

Here we have denoted $v_i^* = v^*(ih)$.

For $n = N$ we get

$$\mathbf{A}_N \vec{V}_{N-1} - \mathbf{C}_N \vec{V}_N = -\vec{F}_N. \tag{3.67}$$

Rewrite the boundary condition (3.23) for $t = t_1$ in the form

$$(2\tilde{\varepsilon}_1\tilde{\varepsilon}_2 + Ft_1^2)\tilde{v}_t(x, t_1) + \tilde{v}^2(x, t_1)\tilde{v}_x^3(x, t_1)t_1^4 = 0, \tag{3.68}$$

where $\tilde{\varepsilon}_1 = \varepsilon_1/t_1$, $\tilde{\varepsilon}_2 = \varepsilon_2/t_1^3$, and $t_1 = N\tau$. Taking advantage of the fact that t_1 is small when solving parabolic equations by the quasi-inversion method (Lattes and Lions, 1970; Samarskii and Vabishchevich, 1990) we omit the term of the higher order of smallness $\tilde{v}^2(x, t_1)\tilde{v}_x^3(x, t_1)t_1^4$. The boundary condition (3.68) takes the form

$$\tilde{v}_t(x, t_1) = 0. \tag{3.69}$$

in numerical solution of the problem, we use the boundary condition for $t = t_1$ in the form (3.69), whereas the condition (3.67) will be used in the matrix form

$$\vec{V}_N = \vec{V}_{N-1}. \tag{3.70}$$

In conclusion we present the solution procedure for the system of matrix equations which is usual for the matrix sweep method (see Samarskii and Nikolaev, 1978):

$$\alpha_{n+1} = (\mathbf{C}_n - \mathbf{A}_n\alpha_n)^{-1}\mathbf{B}_n, \qquad n = 1, \dots, N-1, \quad \alpha_1 = 0;$$

$$\vec{\beta}_{n+1} = (\mathbf{C}_n - \mathbf{A}_n\alpha_n)^{-1}(\mathbf{A}_n\vec{\beta}_n + \vec{F}_n), \qquad n = 1, \dots, N, \quad \vec{\beta}_1 = \vec{F}_0;$$

$$\vec{V}_N = \vec{\beta}_{n+1};$$

$$\vec{V}_n = \alpha_{n+1}\vec{V}_{n+1} + \vec{\beta}_{n+1}, \qquad n = N-1, N-2, \dots, 1, 0.$$

Here $\vec{\beta}_n$ is a vector of dimension $(J-1)$ and α_n is a square matrix of order $(J-1) \times (J-1)$. Write the matrices in the condition (3.67) for the case where the boundary condition for $t = t_1$ has the form (3.69)

$$\mathbf{A}_N = \mathbf{C}_N = \mathbf{E}_{J-1}, \qquad \mathbf{F}_N = 0,$$

where $\mathbf{E}_{J-1}$ is the unit matrix of order $(J-1)$.

A test example. Rewrite equation (3.2) in the form

$$\frac{\partial u}{\partial x}\frac{\mathrm{d}k}{\mathrm{d}x} + \frac{\partial^2 u}{\partial x^2}k = \frac{\partial u}{\partial t}, \qquad x_0 < x < x_1, \quad t > 0,$$

which is a linear ordinary differential equation of the first order for the function $k(x)$. Its solution in the general form can be written as

$$k(x) = \Big(\int \frac{\partial u}{\partial t}\,\mathrm{d}x + C(t)\Big)\Big/\frac{\partial u}{\partial x}. \tag{3.71}$$

Here "the constant of integration" $C(t)$ can be determined, for example, in the following way:

$$C(t) = \Big(k\frac{\partial u}{\partial x}\Big)\Big|_{x=x^*} - \Big(\int \frac{\partial u}{\partial t}\,\mathrm{d}x\Big)\Big|_{x=x^*},$$

where x^* is an arbitrary fixed value of variable x in the interval $[x_0, x_1]$, such as $x^* = x_0$. The value of the derivative $(\partial u/\partial x)(x_0, t)$ is known since it is involved the statement of the coefficient inverse problem. Using the formula (3.71) and specifying the theoretical function $u(x,t)$, we find the theoretical distribution $k(x)$ on condition that the value $k(x_0)$ is known. However, $k(x_0)$ is not involved in the statement of the coefficient inverse problem. Thus, we have to consider other conditions for the problem.

Taking into account the remark on choosing $C(t)$ for an arbitrary value of $x \in [x_0, x_1]$, from formula (3.71) and the condition $k_t = 0$ we obtain the equality

$$u_x \int u_{tt}\,\mathrm{d}x - u_{xt} \int u_t\,\mathrm{d}x = 0,$$

which leads to the condition

$$\frac{\displaystyle\int u_{tt}\,\mathrm{d}x}{\displaystyle\int u_t\,\mathrm{d}x} = \frac{u_{xt}}{u_x}. \tag{3.72}$$

The formula (3.40) implies

$$\begin{vmatrix} u_x & u_{xx} \\ \displaystyle\int_0^t u_x\,\mathrm{d}x & \displaystyle\int_0^t u_{xx}\,\mathrm{d}t \end{vmatrix} \neq 0,$$

which we rewrite in the form

$$\frac{\displaystyle\int_0^t u_{xx}\,\mathrm{d}t}{\int_0^t u_x\,\mathrm{d}t} \neq \frac{u_{xx}}{u_x}. \tag{3.73}$$

The theoretical function $u(x,t)$ should satisfy the conditions (3.72) and (3.73). After introducing the new function $v(x,t)$, the ratio on the right-hand side of the equality

$$v(x,t) = \frac{u_t}{u} \tag{3.74}$$

should depend on the variable t. The theoretical function $u(x,t)$ should satisfy this condition too.

The above-mentioned conditions (3.72)–(3.74) appeared to be sufficient for constructing the theoretical function $u(x,t)$ and determining the theoretical distribution $k(x)$ from the formula (3.71)without using the value $k(x_0)$. For constructing the theoretical distribution $k(x)$ we use the values $(\partial u/\partial x)(x_0,t)$, $u(x_0,t)$, and $u(x,0)$.

Take the theoretical function

$$u(x,t) = x^2 + 40xt + 200t^2, \tag{3.75}$$

which satisfies all the above-mentioned conditions. We determine the theoretical function $k(x)$ from the formula (3.71):

$$k(x) = 10x. \tag{3.76}$$

The coefficient inverse problem for which the theoretical solution is known is formulated as follows. Find a vector function $\{k(x), u(x,t)\}$ satisfying the conditions

$$\frac{\partial}{\partial x}\left(k\frac{\partial u}{\partial x}\right) = \frac{\partial u}{\partial t}, \qquad 1 < x < 2, \quad 0 < t \leq 0.1;$$

$$u(1,t) = 1 + 40t + 200t^2, \qquad u_x(1,t) = 2 + 40t, \qquad 0 < t \leq 10^{-5};$$

$$u(2,t) = 4 + 80t + 200t^2, \qquad u_x(2,t) = 4 + 40t, \qquad 0 < t \leq 10^{-5};$$

$$u(x,0) = x^2, \qquad 1 \leq x \leq 2.$$

Now perform a checking procedure. From the formula (3.76) it follows $k(1) = 10$. We provide this value for solution of initial-boundary problem for equation (3.2) and use it for determining the constant $C(t)$. We obtain $C(t) = 0$. Then we get the same solution $k(x) = 10x$.

We now formulate the auxiliary problem (3.7)–(3.9) obtained after introducing the new function $v(x,t)$:

$$Av_{xx} + Bv_x = Cv_t, \qquad 1 < x < 2, \quad 0 < t \leq 10^{-5};$$

$$v(1,t) = \frac{400t + 40}{200t^2 + 40t + 1}, \quad v_x(1,t) = \frac{-40(200t^2 + 20t + 1)}{(200t^2 + 40t + 1)^2}, \quad 0 < t \leq 10^{-5};$$

$$v(2,t) = \frac{400t + 80}{200t^2 + 80t + 4}, \quad v_x(2,t) = \frac{-40(200t^2 + 40t + 4)}{(200t^2 + 80t + 4)^2}, \quad 0 < t \leq 10^{-5}.$$

This problem has an exact analytic solution

$$v(x,t) = \frac{40x + 400t}{x^2 + 40xt + 200t^2}.$$

The corresponding quasi-inversion problem is

$$\begin{aligned}&-(\tilde{v}^2\tilde{v}_x\tilde{v}_{xt}\tau_1 t)\tilde{v}_{xxxx} + (\tilde{v}^2\tilde{v}_x\tilde{v}_{xxxxt}\tau_1 t)\tilde{v}_x + (F^2 + 4\tilde{\varepsilon}_1\tilde{\varepsilon}_2)t\tilde{v}_{tt}\\ &+ 2F(2F - \tilde{\varepsilon}_1)\tilde{v}_t + 6F\tilde{v}^2\tilde{v}_x^3 t^2 = 0, \qquad 1 < x < 2, \quad 0 < t \leq 10^{-5};\end{aligned}$$

$$\tilde{v}(1,t) = \frac{400t + 40}{200t^2 + 40t + 1}, \quad \tilde{v}_x(1,t) = \frac{-40(200t^2 + 20t + 1)}{(200t^2 + 40t + 1)^2}, \quad 0 < t \leq 10^{-5};$$

$$\tilde{v}(2,t) = \frac{400t+80}{200t^2+80t+4}, \quad \tilde{v}_x(2,t) = \frac{-40(200t^2+40t+4)}{(200t^2+80t+4)^2}, \quad 0 < t \leq 10^{-5}.$$

$$\tilde{v}(x,0) = v^*(x), \quad 1 \leq x \leq 2; \qquad \tilde{v}_t(x,t_1) = 0, \quad 1 \leq x \leq 2.$$

Here we have used the designations $F = 2\tilde{v}_{xx}/x + 6\tilde{v}_x/x^2$ and $f = 2/x$.

Now consider the **algorithm of choosing the parameter** τ_1 in the process of calculations. Since $\tau_1 = t - \bar{t}$ and value $\bar{t}$ is determined by the mean-value theorem applied to the integrals in the coefficients of equation (3.11) (see Section 3.3), we have $\tau_1 = \tau_1(x,t)$. For any value of t, when x is fixed, finding τ_1 is reduced to determining the corresponding value of $\bar{t}$.

We analyze the algorithm of numerical solution of the problem. The function $\tau_1(x,t)$ entering the coefficients of equation (3.56) also enters the coefficients α_{n+1} and β_{n+1} of the sweep method. Knowing these coefficients we find the unknown vector function $\vec{V}_n$, i.e., determine the values of the mesh function $\tilde{v}_i^n$. These values are not known beforehand. They are calculated using the sweep method which is reverse in time; therefore, they cannot be used in the beginning to find the values of $\bar{t}$. Consequently, the values of the function $\tau_1(x,t)$ also cannot be specified before the calculation of the vector function $\vec{V}_n$.

We propose several methods of specifying the function $\tau_1(x,t)$.

In **the first method**, calculations are performed on a small interval of time t. For this reason we shall assume that $\bar{t} = 0$. By the mean-value theorem, $0 < \bar{t} < t$ and therefore the foregoing assumption has an error which can be considered small. In this case we obtain $\tau_1 = t$.

The second method consists of several stages. On the first stage, applying the numerical interpretation of the mean-value theorem, for every value of t we find the corresponding value $\bar{t}$ using the values of the function specified on the boundary of the solution domain, $\tilde{v}(x_0,t)$ and $\tilde{v}(x_1,t)$. As the result, we find the values $\tau_1(x_0,t)$ and $\tau_1(x_1,t)$. After that, guessing the properties of the sought function $\tilde{v}(x,t)$ from physical considerations, we perform (possibly, linear) interpolation of the values of τ_1 with respect to x for every value of t. The obtained values will be used further in the calculation process.

Finally, **the third method** of specifying the auxiliary function $\tau_1(x,t)$ may be presented as a method of its successive refinement in an iterative process. This will certainly lead to the considerable increase in the amount of calculation.

We shall use the first method for solving a specific problem. Conclusions about its adequacy will be made on obtaining the results of the computational experiment. In this approach, the equation (3.53) of the quasi-inversion method takes the form

$$-(\tilde{v}^2\tilde{v}_x\tilde{v}_{xt}t^2)\tilde{v}_{xxxx} + (\tilde{v}^2\tilde{v}_x\tilde{v}_{xxxxt}t^2)\tilde{v}_x + (F^2 + 4\tilde{\varepsilon}_1\tilde{\varepsilon}_2)t\tilde{v}_{tt}$$
$$+ (2F - \tilde{\varepsilon}_1)2F\tilde{v}_t + 6F\tilde{v}^2\tilde{v}_x^3t^2 = 0.$$

3.8. RESULTS OF SOLUTION OF A TEST EXAMPLE PROBLEM

The method of solving a problem with non-classical set of additional conditions (the problem of continuation of a solution of a parabolic equation), to which the solution of the coefficient inverse problem is finally reduced, is described in detail in Danilaev (1989) and Lattes and Lions (1967). In this case, in determining the auxiliary function $\tilde{v}(x,t)$ the results of these studies proved to be valid. The methods of choosing the regularization parameters and the instant of time at which it is necessary to estimate the calculation results also proved to be valid. The influence of the choice of the mesh length on the numerical solution behavior was the same (as described in Lattes and Lions, 1967).

The results of numerical calculations for the test example are presented in Figures 3.1 and 3.2. The characteristic behavior of the numerical solution of the quasi-inversion problem is shown in Figure 3.1, where curve *1* represents the results of numerical solution of the quasi-inversion problem corresponding to the test example, and curve *2* represents the exact analytic values corresponding to the function $\tilde{v}(x,t)$. The calculation was done for the mesh domain with 10 cells with respect to x and 100 cells with respect to t. The mesh lengths were $h = 10^{-1}$ for x and $\tau = 10^{-7}$ for t. The values of the regularization parameters in the calculation illustrated by the given graph were $\varepsilon_1 = 10^{-6}$ and $\varepsilon_2 = 10^{-5}$. The initial estimate was constructed as a result of linear interpolation of the function values given on the boundary of the solution domain. The number of iterations in the calculation process (from 1 to 7) depended on the choice of the regularization parameters and the mesh lengths.

In Figure 3.2, the values of the coefficient $k(x)$ found as a result of numerical solution of the problem (dashed curve *2*) are compared with its exact values (solid curve *1*).

After solving the quasi-inversion problem, the problem of determining the coefficient $k(x)$ was reduced to solution of the following system of linear

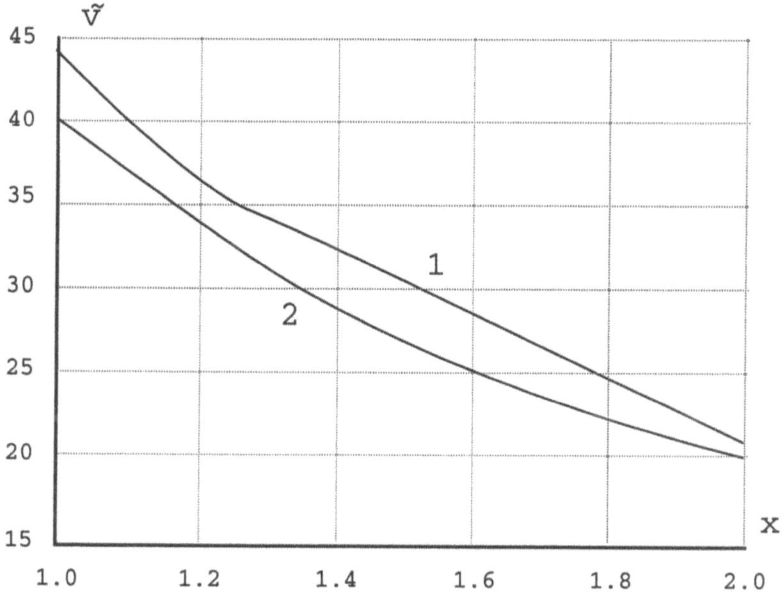

Figure 3.1

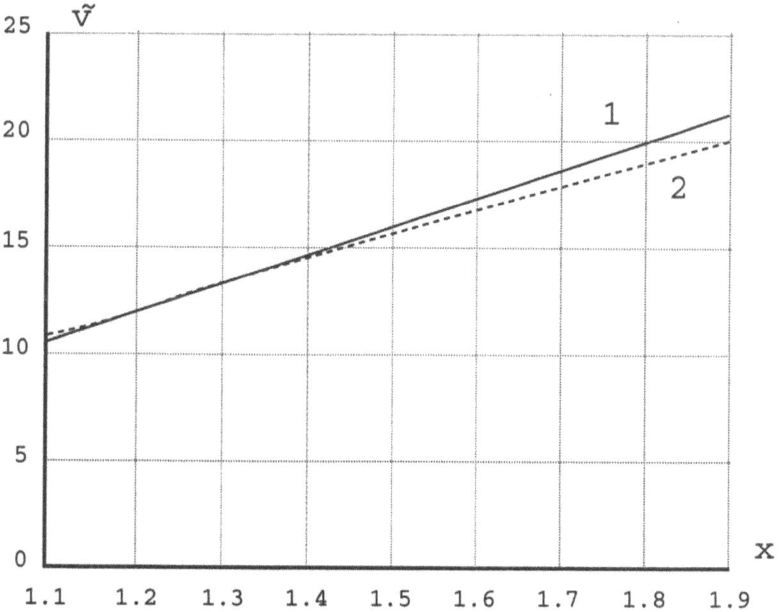

Figure 3.2

algebraic equations

$$\begin{cases} k_x u_x + k u_{xx} = u_t, \\ k_x \displaystyle\int_0^t u_x \, \mathrm{d}t + k \int_0^t u_{xx} \, \mathrm{d}t = \int_0^t u_t \, \mathrm{d}t, \end{cases}$$

whose the coefficients were expressed beforehand in terms of the new auxiliary function $\tilde{v}(x, t$. The problem was solved on the small half-open interval $(0, t_1]$, however, the above system of linear algebraic equations appeared to be well-conditioned. Its solution could be found without the regularization procedure. Having solved the system for the fixed value $x = x_k$, we then find the coefficient $k(x)$ from the formula

$$k(x) = \frac{1}{u_x(x)} \Big[k(x_k) u(x_k) + \int_{x_k}^{x} u_t \, \mathrm{d}x \Big].$$

The influence of the perturbed boundary conditions on the stability of the problem solution was also studied. The error $\delta = ax + \sin bt$ was added to the function and its derivative on the boundary of the solution domain: $u(x, t) = x^2 + 40xt + 200t^2 + \delta$. The specified constants a and b determined the value of the relative error which was up to 5 % in the norm of C^2. This error leads to a small error in the solution and had almost no influence on the result. The graphs illustrating the solution of the perturbed problem coincide with the curves shown in Figure 3.1 and 3.2.

Chapter 4.

Modification of the method of determining the coefficient of the leading terms in an equation

4.1. MODIFICATION METHOD

In the previous chapter we considered the initial-boundary value problem of determination of the vector function $\{k(x), u(x,t)\}$ which satisfies the equation

$$(ku_x)_x = u_t$$

and an overdetermined set of additional conditions. After introducing the new unknown function $v(x,t) = u_t/u$ and eliminating the unknown coefficient from equation (3.1), the problem was reduced to solution of the equation

$$Av_{xx} + Bv_x = Cv_t. \tag{4.1}$$

In Section 3.3 we analyzed the coefficients of equation (4.1) assuming that its solution is sought on a small time interval. Under this assumption, the coefficients of equation were simplified and took the form

$$A = -\frac{1}{2}\bar{v}\bar{v}_x t^2, \qquad B = \frac{1}{2}\bar{v}\bar{v}_{xx} t^2, \qquad C = \frac{1}{2}[f\bar{v}_{xx} + (f^2 - f_x)\bar{v}_x]t^2.$$

Then we simplified the equation of the corresponding quasi-inversion problem.

Consider another approach. Substitute the simplified values of the coefficients A, B and C into equation (4.1) and analyze the obtained result:

$$-\bar{v}\bar{v}_x v_{xx}+\bar{v}\bar{v}_{xx}v_x = [f\bar{v}_{xx}+(f^2-f_x)\bar{v}_x]v_t, \qquad [f\bar{v}_{xx}+(f^2-f_x)\bar{v}_x] = 0 \quad (4.2)$$

$$\frac{\bar{v}_{xx}}{\bar{v}_x} = \frac{f_x}{f} - f.$$

Equation (4.2) is an ordinary differential equation for the function $\bar{v}(x,t)$ depending on t as a parameter. It is obvious that $v_t \neq 0$ since $u_t \neq 0$ by assumption.

For the function $\bar{v}(x,t)$, we formulate the problem

$$f v_{xx} + (f^2 - f_x)v_x = 0, \qquad x_0 < x < x_1, \quad -\varepsilon < t \le t_1,$$

$$v(x_0,t) = \mu_0(t), \qquad v_x(x_0,t) = \eta_0(t), \qquad -\varepsilon < t \le t_1,$$

$$f(x) = \varphi_x/\varphi, \qquad \varphi(x) = u(x,0), \qquad x_0 < x < x_1. \quad (4.3)$$

Integrate the ordinary differential equation of the problem (4.3):

$$\frac{v_{xx}}{v_x} = \frac{f_x}{f} - f, \qquad (\ln v_x)_x = (\ln f)_x - f,$$

$$\int_{x_0}^{x} (\ln v_x)_x \, dx = \int_{x_0}^{x} (\ln f)_x \, dx - \int_{x_0}^{x} f(x)\, dx,$$

$$\ln\frac{v_x(x,t)}{v_x(x_0,t)} = \ln\frac{f(x)}{f(x_0)} - \int_{x_0}^{x} f(x)\, dx,$$

$$v_x(x,t) = \eta_0(t)\,\frac{f(x)}{f(x_0)}\exp\Big[-\int_{x_0}^{x} f(x)\, dx\Big].$$

Calculate the integral in the right-hand side of this equality:

$$\int_{x_0}^{x} f(x)\, dx = \int_{x_0}^{x} (\ln\varphi)_x\, dx = \ln\Big[\frac{\varphi(x)}{\varphi(x_0)}\Big], \quad \exp\Big[-\int_{x_0}^{x} f(x)\, dx\Big] = \frac{\varphi(x)}{\varphi(x_0)}.$$

We obtain

$$v_x(x,t) = \eta_0(t)\,\frac{f(x)}{f(x_0)}\,\frac{\varphi(x_0)}{\varphi(x)} = \eta_0(t)\Big\{\frac{\varphi_x(x)}{\varphi(x)}\,\frac{\varphi(x_0)}{\varphi(x)}\Big/\frac{\varphi_x(x_0)}{\varphi(x_0)}\Big\},$$

$$v_x(x,t) = \frac{a(t)\varphi_x(x)}{\varphi^2(x)}, \quad (4.4)$$

where $a(t) = \eta_0(t)\varphi^2(x_0)/\varphi_x(x_0)$. Integrate equality (4.4):

$$\int_{x_0}^{x} v_x(x,t)\,\mathrm{d}x = a(t)\int_{x_0}^{x} -\Big(\frac{1}{\varphi(x)}\Big)'_x\,\mathrm{d}x,$$

$$v(x,t) - \mu_0(t) = -a(t)\Big[\frac{1}{\varphi(x)} - \frac{1}{\varphi(x_0)}\Big]. \tag{4.5}$$

As follows from the introduced notation, the equality $v(x,t) = (\ln u(x,t))_t$ is valid. Substituting this equality into (4.5), we obtain

$$(\ln u(x,t))_t = \mu_0(t) + a(t)\Big[\frac{1}{\varphi(x)} - \frac{1}{\varphi(x_0)}\Big],$$

$$\int_0^t (\ln u(x,t))_t\,\mathrm{d}t = \int_0^t \mu_0(t)\,\mathrm{d}t + \Big[\frac{1}{\varphi(x_0)} - \frac{1}{\varphi(x)}\Big]\int_0^t a(t)\,\mathrm{d}t,$$

$$\ln\frac{u(x,t)}{\varphi(x)} = \int_0^t \mu_0(t)\,\mathrm{d}t + \Big[\frac{1}{\varphi(x_0)} - \frac{1}{\varphi(x)}\Big]\frac{\varphi^2(x_0)}{\varphi_x(x_0)}\int_0^t \eta_0(t)\,\mathrm{d}t,$$

We now turn to the statement of the problem (3.3) considered for equation (3.2) in Section 3.1. Use the relation between the functions $u(x,t)$ and $v(x,t)$ and determine the relations between their boundary conditions. Transform the expression for the derivative

$$v_x(x,t) = (\ln u(x,t))_{tx} = (\ln u(x,t))_{xt} = \Big(\frac{u_x(x,t)}{u(x,t)}\Big)'_t.$$

Determine the relation between the boundary conditions

$$u_x(x_0,t) = h_0(t), \qquad u(x_0,t) = g_0(t), \qquad v_x(x_0,t) = \eta_0(t).$$

We obtain

$$\eta_0(t) = \Big(\frac{u_x(x_0,t)}{u(x_0,t)}\Big)_t = \Big(\frac{h_0(t)}{g_0(t)}\Big)_t, \mu_0(t) = (\ln g_0(t))_t,$$

$$\int_0^t \eta_0(t)\,\mathrm{d}t = \int_0^t \Big(\frac{h_0(t)}{g_0(t)}\Big)_t\,\mathrm{d}t = \frac{h_0(t)}{g_0(t)}\Big|_0^t = \frac{h_0(t)}{g_0(t)} - \frac{h_0(0)}{g_0(0)} = \frac{h_0(t)}{g_0(t)} - \frac{\varphi_x(x_0)}{\varphi(x_0)},$$

$$\int_0^t \mu_0(t)\,\mathrm{d}t = \int_0^t (\ln g_0(t))_t\,\mathrm{d}t = \ln\frac{g_0(t)}{g_0(0)}.$$

Here we used the usual condition of concordance of the boundary and initial conditions. Finally, we obtain

$$\ln\frac{u(x,t)}{\varphi(x)} = \ln\frac{g_0(t)}{\varphi(x_0)} + \Big(\frac{1}{\varphi(x_0)} - \frac{1}{\varphi(x)}\Big)\frac{\varphi^2(x_0)}{\varphi_x(x_0)}\Big(\frac{h_0(t)}{g_0(t)} - \frac{\varphi_x(x_0)}{\varphi(x_0)}\Big). \tag{4.6}$$

From the formula (4.6) we obtain

$$u(x,t) = \frac{\varphi(x)}{\varphi(x_0)}\, g_0(t) \exp[\chi(x,x_0)\phi(t,x_0)], \tag{4.7}$$

where

$$\chi(x,x_0) = \Big(\frac{1}{\varphi(x_0)} - \frac{1}{\varphi(x)}\Big)\frac{\varphi^2(x_0)}{\varphi_x(x_0)},$$

$$\begin{aligned}\phi(t,x_0) &= \Big(\frac{h_0(t)}{g_0(t)} - \frac{\varphi_x(x_0)}{\varphi(x_0)}\Big) = \Big(\frac{u_x(x_0,t)}{u(x_0,t)} - \frac{\varphi_x(x_0)}{\varphi(x_0)}\Big)\\ &= (\ln u)_x(x_0,t) - (\ln\varphi)_x(x_0,t) = \Big(\ln\frac{u}{\varphi}\Big)_x(x_0,t).\end{aligned}$$

It is obvious that $\lim_{t\to\infty}\phi(t,x_0) = 0$.

Write the expression $\exp[\chi(x,x_0)\phi(t,x_0)]$ in the form of the series expansion in powers of $\phi(t,x_0)$ in the neighbourhood of the point $\phi = 0$

$$\exp[\chi(x,x_0)\phi(t,x_0)] = 1 + \chi\phi + \frac{(\chi\phi)^2}{2} + \ldots + \frac{(\chi\phi)^n}{n!} + \frac{(\chi\phi)^{n+1}}{(n+1)!}e^c, \tag{4.8}$$

where

$$c = \chi(x,x_0)\phi(\theta t,x_0) = \chi(x,x_0)\Big(\frac{u_x(x_0,\theta t)}{u(x_0,\theta t)} - \frac{\varphi_x(x_0)}{\varphi(x_0)}\Big), \quad 0<\theta<1.$$

For small values of t the function $\phi(t,x_0)$ is small, since for small values of t the values of the functions $u(x,t)$ and $\varphi(x)$ are close and therefore $\ln(u/\varphi)$ tends to zero. For this reason, in the expansion (4.8) we use only two terms and substitute it into the formula (4.7)

$$\begin{aligned}u(x,t) &= \frac{\varphi(x)}{\varphi(x_0)}\, g_0(t)[1 + \chi(x,x_0)\phi(t,x_0) + R_2(x,t,x_0)]\\ &= \frac{\varphi(x)}{\varphi(x_0)}\, g_0(t)\Big[1 + \Big(\frac{1}{\varphi(x_0)} - \frac{1}{\varphi(x)}\Big)\frac{\varphi^2(x_0)}{\varphi_x(x_0)}\Big(\frac{h_0(t)}{g_0(t)} - \frac{\varphi_x(x_0)}{\varphi(x_0)}\Big)\\ &\qquad + R_2(x,t,x_0)\Big],\end{aligned} \tag{4.9}$$

where $R_2(x,t,x_0)$ is a remainder term. Simplify the obtained result (4.9):

$$u(x,t) = g_0(t) + (\varphi(x) - \varphi(x_0))\frac{h_0(t)}{\varphi_x(x_0)} + \frac{\varphi(x)}{\varphi(x_0)}\, g_0(t)\, R_2(x,t,x_0). \tag{4.10}$$

The formula (4.10) can be formally written in the form which is a special case of the Cauchy theorem

$$u(x,t) - u(x_0,t) = (\varphi(x) - \varphi(x_0))\frac{u_x(x_0,t)}{\varphi_x(x_0,t)} + O(\chi^2\phi^2) \tag{4.11}$$

or

$$\frac{u(x,t) - u(x_0,t)}{\varphi(x) - \varphi(x_0)} = \frac{u_x(x_0,t)}{\varphi_x(x_0,t)}. \tag{4.12}$$

Generalizing the obtained results we formulate two theorems.

Theorem 4.1. *Suppose that on a small time half-interval* $(0,t_1]$ *a solution* $\{k(x),u(x,t)\}$ *of the following coefficient inverse problem is sought:*

$$\begin{gathered}(k(x)u_x)_x = u_t, \qquad x_0 < x < x_1, \quad -\varepsilon < t \le t_1,\\ u(x_0,t) = g_0(t), \qquad u_x(x_0,t) = h_0(t), \qquad -\varepsilon < t \le t_1,\\ u(x,0) = \varphi(x), \qquad x_0 < x < x_1.\end{gathered} \tag{4.13}$$

Then the component $u(x,t)$ *of sought vector function can be determined to an accuracy of* $O(\chi^2\phi^2)$ *by the formula (4.10) which can be formally transformed to the form (4.11) or (4.12).*

Theorem 4.2. *Suppose that it is required to find a solution* $\{k(x),u(x,t)\}$ *of the coefficient inverse problem (4.13) defined on a small time half-interval* $(0,t_1]$. *Then the component* $u(x,t)$ *of the solution can be written within the accuracy of* $O(\chi\phi)$ *in the form*

$$u(x,t) = \frac{\varphi(x)}{\varphi(x_0)}\, g_0(t) + O(\chi\phi),$$

i.e. $u(x,t)$ *admits separation of variables of the form*

$$u(x,t) = U(x)T(t), \qquad U(x) = \frac{\varphi(x)}{\varphi(x_0)}, \qquad T(t) = g_0(t).$$

Remark 4.1. Consider the behavior of the obtained solution (4.10) in the neighborhood of the point $x = x_0$. For $x = x_0$ the formula (4.10) becomes an identity:

$$u(x_0,t) \equiv u(x_0,t) + u(x_0,t)R_2(x_0,t,x_0),$$

where $R_2(x_0,t,x_0) \equiv 0$, since in this case $\chi(x_0,x_0) = 0$ and $e^c = e^0 = 1$, the function $\phi(t,x_0)$ having no singularities.

Remark 4.2. After the component $u(x,t)$ of of the sought vector-function $\{k(x), u(x,t)\}$ is found, the other component, i.e., the function $k(x)$, can be determined from the same formulas as in the previous chapter, for example,

$$k(x) = \left[u_t \int_0^t u_x \,\mathrm{d}t - u_x(u-\varphi)\right] \Big/ \left[u_{xx} \int_0^t u_x \,\mathrm{d}t - u_x \int_0^t u_{xx} \,\mathrm{d}t\right].$$

Theorem 4.3. *Suppose that it is required to find a solution $\{k(x), u(x,t)\}$ of the coefficient inverse problem (4.13) defined on a small half-interval $(0, t_1)$, and the function $\varphi(x)$ in the additional conditions of the problem is twice differentiable on $[x_0, x_1]$. Then the component $u(x,t)$ of the sought vector function can be determined from the formula (4.10) in the neighbourhood of the point x_0 within the accuracy of not higher than $O((x-x_0)^2)$.*

Proof. Apply Lagrange's theorem to the difference $\varphi(x) - \varphi(x_0)$

$$\varphi(x) - \varphi(x_0) = \varphi'_x((1-\theta_1)x_0 + \theta_1 x)(x - x_0),$$

where $0 < \theta_1 < 1$, $x_0 + \theta_1(x - x_0) = (1-\theta_1)x_0 + \theta_1 x$. Then we obtain

$$\begin{aligned} \chi(x, x_0) &= \frac{\varphi'_x((1-\theta_1)x_0 + \theta_1 x)(x-x_0)}{\varphi(x)\varphi(x_0)} \frac{\varphi^2(x_0)}{\varphi'(x_0)} \\ &= \frac{\varphi'_x((1-\theta_1)x_0 + \theta_1 x)}{\varphi'(x_0)} \frac{\varphi(x_0)}{\varphi(x)} (x - x_0). \end{aligned} \tag{4.14}$$

As was proved above, the function $\phi(t, x_0)$ tends to zero as $t \to 0$. We also wrote the function $\exp(\chi(x,x_0)\phi(t,x_0))$ in powers ϕ in the neighbourhood of the point $\phi = 0$. By virtue of (4.14), the remainder term of this expansion is

$$\begin{aligned} R_2(x,t,x_0) &= \frac{1}{2}\chi^2(x,x_0)\phi^2(t,x_0)\mathrm{e}^{c_1} \\ &= \frac{1}{2}\left(\frac{\varphi'_x((1-\theta_1)x_0 + \theta_1 x)}{\varphi'(x_0)} \frac{\varphi(x_0)}{\varphi(x)}\right)^2 (x-x_0)^2 \left(\left(\ln\frac{u}{\varphi}\right)'_x(x_0, t)\right)^2 \mathrm{e}^{c_1}, \end{aligned}$$

where

$$c_1 = \frac{\varphi'_x((1-\theta_1)x_0 + \theta_1 x)}{\varphi'(x_0)} \frac{\varphi(x_0)}{\varphi(x)} (x - x_x)\left(\left(\ln\frac{u}{\varphi}\right)'_x(x_0, \theta t)\right),$$

$$0 < \theta, \theta_1 < 1.$$

Obviously,

$$\lim_{t\to 0} c_1 = 0.$$

Denote $\zeta(t, x_0) = u_x(x_0, t)/u(x_0, t)$. Take the Maclaurin series expansion of $\zeta(t, x_0)$ in powers of t. We obtain the following equality:

$$\begin{aligned}
\zeta(t, x_0) &= \zeta(0, x_0) + \zeta_t'(0, x_0)t + R_2^* \\
&= \frac{\varphi_x'(x_0)}{\varphi(x_0)} + \left[\frac{u_{xt}(x_0, 0)}{\varphi(x_0)} - \frac{\varphi_x'(x_0)}{\varphi(x_0)}\frac{u_t(x_0, 0)}{\varphi(x_0)}\right]t + R_2^* \\
&= \frac{\varphi_x'(x_0)}{\varphi(x_0)} + \left[\frac{h_{0t}'(0)}{\varphi(x_0)} - \frac{\varphi_x'(x_0)}{\varphi(x_0)}\frac{g_{0t}'(0)}{\varphi(x_0)}\right]t + R_2^* \\
&= \frac{\varphi_x'(x_0)}{\varphi(x_0)} + \frac{h_{0t}'(0)g_0(0) - h_0(0)g_{0t}'(0)}{g_0^2(0)}t + R_2^* \\
&= \frac{\varphi_x'(x_0)}{\varphi(x_0)} + \left(\frac{h_0}{g_0}\right)_t' t + R_2^*.
\end{aligned}$$

Consequently,

$$\phi(t, x_0) = \zeta(t, x_0) - \frac{\varphi_x'(x_0)}{\varphi(x_0)} = \left(\frac{h_0}{g_0}\right)_t'(0)t + R_2^*,$$

$\phi(t, x_0) = O(t)$, $R_2^*(x, t, x_0) = O((x - x_0)^2 t^2)$ which proves the theorem. □

4.2. DEFINING A COEFFICIENT

In Section 3.5 we deduced the formulas for calculating $k(x)$ and the derivative $k_x(x)$. We shall show that these formulas can be obtained in another way and written in a different form.

Theorem 4.4. *The formulas in Section 3.5 for calculation the unknown coefficient $k(x)$ in equation (4.13) and its derivative $k_x(x)$ can be found as a result of solution of a system of linear algebraic equations. The first of these equations is the original equation of the problem, and the second can be obtained from the first either by integration with respect to t (any number of times) or by differentiation with respect to t (any number of times).*

Proof. The following formulas for calculating the unknown coefficient $k(x)$ and its derivative were deduced above:

$$k(x) = \left(u_x \int_0^t u_t\,\mathrm{d}t - u_t \int_0^t u_x\,\mathrm{d}t\right) \Big/ \left(u_x \int_0^t u_{xx}\,\mathrm{d}t - u_{xx} \int_0^t u_x\,\mathrm{d}t\right), \quad (4.15)$$

$$k_x(x) = -\left(u_{xx}\int_0^t u_t\,\mathrm{d}t - u_t\int_0^t u_{xx}\,\mathrm{d}t\right)\Big/\left(u_x\int_0^t u_{xx}\,\mathrm{d}t - u_{xx}\int_0^t u_x\,\mathrm{d}t\right). \tag{4.16}$$

Formulas (4.15), (4.16) can be written in the following form:

$$k(x) = \begin{vmatrix} u_x & u_t \\ \int_0^t u_x\,\mathrm{d}t & \int_0^t u_t\,\mathrm{d}t \end{vmatrix} \Big/ \begin{vmatrix} u_x & u_{xx} \\ \int_0^t u_x\,\mathrm{d}t & \int_0^t u_{xx}\,\mathrm{d}t \end{vmatrix},$$

$$k_x(x) = -\begin{vmatrix} u_{xx} & u_t \\ \int_0^t u_{xx}\,\mathrm{d}t & \int_0^t u_t\,\mathrm{d}t \end{vmatrix} \Big/ \begin{vmatrix} u_x & u_{xx} \\ \int_0^t u_x\,\mathrm{d}t & \int_0^t u_{xx}\,\mathrm{d}t \end{vmatrix}. \tag{4.17}$$

From (4.17) it follows that the unknown functions $k(x)$ and $k_x(x)$ are the solution of the system of two linear algebraic equations

$$k_x u_x + k u_{xx} = u_t,$$

$$k_x\int_0^t u_x\,\mathrm{d}t + k\int_0^t u_{xx}\,\mathrm{d}t = \int_0^t u_t\,\mathrm{d}t. \tag{4.18}$$

□

Consider formulas (4.15), (4.16) under the assumption that the upper limit of integration, t, is small. The numerator and denominator of these fractions tend to zero as t. Applying the L'Hospital's rule, we obtain the equations

$$\begin{aligned} k(x) &= \left(u_{xt}\int_0^t u_t\,\mathrm{d}t - u_{tt}\int_0^t u_x\,\mathrm{d}t\right)\Big/\left(u_{xt}\int_0^t u_{xx}\,\mathrm{d}t - u_{xxt}\int_0^t u_x\,\mathrm{d}t\right) \\ &= \begin{vmatrix} u_{xt} & u_{tt} \\ \int_0^t u_x\,\mathrm{d}t & \int_0^t u_t\,\mathrm{d}t \end{vmatrix} \Big/ \begin{vmatrix} u_{xt} & u_{xxt} \\ \int_0^t u_x\,\mathrm{d}t & \int_0^t u_{xx}\,\mathrm{d}t \end{vmatrix}, \end{aligned} \tag{4.19}$$

$$\begin{aligned} k_x(x) &= \left(u_{tt}\int_0^t u_{xx}\,\mathrm{d}t - u_{xxt}\int_0^t u_t\,\mathrm{d}t\right)\Big/\left(u_{xt}\int_0^t u_{xx}\,\mathrm{d}t - u_{xxt}\int_0^t u_x\,\mathrm{d}t\right) \\ &= \begin{vmatrix} u_{tt} & u_{xxt} \\ \int_0^t u_t\,\mathrm{d}t & \int_0^t u_{xx}\,\mathrm{d}t \end{vmatrix} \Big/ \begin{vmatrix} u_{xt} & u_{xxt} \\ \int_0^t u_x\,\mathrm{d}t & \int_0^t u_{xx}\,\mathrm{d}t \end{vmatrix}. \end{aligned} \tag{4.20}$$

From (4.19), (4.20) it follows that the unknown functions $k(x)$, $k_x(x)$ can be viewed as a solution of the system of two linear algebraic equations

$$k_x u_{xt} + k u_{xxt} = u_{tt},$$

$$k_x \int_0^t u_x \, \mathrm{dt} + k \int_0^t u_{xx} \, \mathrm{dt} = \int_0^t u_t \, \mathrm{dt}. \tag{4.21}$$

We now prove that the unknown functions $k(x)$ and $k_x(x)$ can be also viewed as a solution of the following system of equations:

$$\begin{aligned} k_x u_x + k u_{xx} &= u_t, \\ k_x u_{xt} + k u_{xxt} &= u_{tt}. \end{aligned} \tag{4.22}$$

To this end, we formulate the following lemma:

Lemma 4.1. *The solutions of the systems of algebraic equations* (4.18), (4.21) *coincide.*

Proof. Equate the left-hand sides of formulas (4.15) and (4.19) with each other. We prove that the obtained equality

$$\begin{aligned} &\begin{vmatrix} u_{xt} & u_{tt} \\ \int_0^t u_x \, \mathrm{dt} & \int_0^t u_t \, \mathrm{dt} \end{vmatrix} \Big/ \begin{vmatrix} u_x & u_{xx} \\ \int_0^t u_x \, \mathrm{dt} & \int_0^t u_{xx} \, \mathrm{dt} \end{vmatrix} \\ &= \begin{vmatrix} u_{xt} & u_{tt} \\ \int_0^t u_x \, \mathrm{dt} & \int_0^t u_t \, \mathrm{dt} \end{vmatrix} \Big/ \begin{vmatrix} u_{xt} & u_{xxt} \\ \int_0^t u_x \, \mathrm{dt} & \int_0^t u_{xx} \, \mathrm{dt} \end{vmatrix} \end{aligned} \tag{4.23}$$

is an identity. Reducing equality (4.23) to a common denominator, calculating the determinants and transform the result, we have

$$\begin{vmatrix} u_x & u_{xx} & u_t \\ u_{xt} & u_{xxt} & u_{tt} \\ \int_0^t u_x \, \mathrm{dt} & \int_0^t u_{xx} \, \mathrm{dt} & \int_0^t u_t \, \mathrm{dt} \end{vmatrix} = 0. \tag{4.24}$$

We now turn to equation (3.6). It is obvious that it can be written in the form (4.24). Consequently, equality (4.23) is an identity, which proves the lemma. □

We return to the proof of Theorem 4.4. The solutions of the systems of algebraic equations (4.18), (4.21) coincide. Since these solutions turn the system equations into identities, they also are solutions of (4.22). This completes the proof of Theorem 4.4.

Theorem 4.5. *Formulas* (4.17) *of calculating the unknown coefficient and its derivative can be used only if the function* $u(x, t)$ *is determined within* *the accuracy higher than $O(b^2\xi^2)$ (the designations are the same as before). Using formula (4.17) for small t is equivalent to solving an ill-conditioned system of linear algebraic equations.*

Proof. Assume that the component $u(x,t)$ of the sought solution of the problem (4.3) is determined within the accuracy of $O(b^2\xi^2)$. Write this function according to the formula (4.10)

$$u(x,t) = g_0(t) + (\varphi(x) - \varphi(x_0))\frac{h_0(t)}{\varphi'_x(x_0)} + O(b^2\xi^2). \tag{4.25}$$

Using (4.25), calculate the determinant in the denominators of formulas (4.17)

$$\begin{vmatrix} u_x & u_{xx} \\ \int_0^t u_x\,\mathrm{d}t & \int_0^t u_{xx}\,\mathrm{d}t \end{vmatrix} = \begin{vmatrix} \varphi'_x(x)\dfrac{h_0(t)}{\varphi'_x(x_0)} & \varphi'_{xx}(x)\dfrac{h_0(t)}{\varphi'_x(x_0)} \\ \dfrac{\varphi'_x(x)}{\varphi'_x(x_0)}\displaystyle\int_0^t h_0(t)\,\mathrm{d}t & \dfrac{\varphi'_{xx}(x)}{\varphi'_x(x_0)}\displaystyle\int_0^t h_0(t)\,\mathrm{d}t \end{vmatrix} = 0.$$

Next, calculate the determinants in the numerators of formulas (4.17). Under the same conditions, we find that they are nonzero, but tend to zero as t becomes small. For example,

$$\begin{vmatrix} u_x & u_{xx} \\ \int_0^t u_x\,\mathrm{d}t & \int_0^t u_{xx}\,\mathrm{d}t \end{vmatrix} = \frac{\varphi'_x(x)}{\varphi'_x(x_0)}\begin{vmatrix} h_0(t) & g_{0t}(t) \\ \int_0^t h_0(t)\,\mathrm{d}t & \int_0^t g_{0t}(t)\,\mathrm{d}t \end{vmatrix}$$
$$- \frac{\varphi'_x(x)(\varphi(x) - \varphi(x_0))}{\varphi'^2_x(x_0)}\begin{vmatrix} h_0(t) & h_{0t}(t) \\ \int_0^t h_0(t)\,\mathrm{d}t & \int_0^t h_{0t}(t)\,\mathrm{d}t \end{vmatrix}.$$

Thus, if the function $u(x,t)$ is calculated within the accuracy of $O(b^2\xi^2)$, then it is impossible to use formulas (4.17) for calculating the unknown coefficient and its derivative because the denominators of the corresponding formulas vanish.

We now prove that, in general, formulas (4.17) can be used for determining the functions $k(x)$ and $k_x(x)$. We calculate the determinants in the denominators of these formulas applying the mean-value theorem to the integrals therein:

$$\begin{vmatrix} u_x & u_{xx} \\ \int_0^t u_x\,\mathrm{d}t & \int_0^t u_{xx}\,\mathrm{d}t \end{vmatrix} = \begin{vmatrix} u_x(x,t) & u_{xx}(x,t) \\ u_x(x,\bar{t})t & u_{xx}(x,\bar{\bar{t}})t \end{vmatrix} = t\begin{vmatrix} u_x(x,t) & u_{xx}(x,t) \\ u_x(x,\bar{t}) & u_{xx}(x,\bar{\bar{t}}) \end{vmatrix} \neq 0,$$

where $0 < \bar{t}, \bar{\bar{t}} < t$, $\bar{t}$ and $\bar{\bar{t}}$ are constants (in general, $\bar{t} \neq \bar{\bar{t}}$). Then we transform the numerators in (4.17) in a similar way. As a result of these operations, we conclude that after finding $u(x,t)$ these formulas can be considered as a solution of the system of linear algebraic equations

$$u_x(x,t)k_x(x) + u_{xx}(x,t)k(x) = u_t(x,t),$$

$$u_x(x,\bar{t})k_x(x) + u_{xx}(x,\bar{\bar{t}})k(x) = u_t(x,\bar{\bar{\bar{t}}}), \tag{4.26}$$

where $\bar{\bar{\bar{t}}}$ is also a constant, and $0 < \bar{\bar{\bar{t}}} < t$. The system (4.26) can be simplified using the fact that the interval $[0,t]$ is small and setting $\bar{t} = \bar{\bar{t}} = \bar{\bar{\bar{t}}} = t^* =$ const. Then we get the system of linear algebraic equations for the unknown functions $k(x)$ and $k_x(x)$

$$u_x(x,t)k_x(x) + u_{xx}(x,t)k(x) = u_t(x,t),$$

$$u_x(x,t^*)k_x(x) + u_{xx}(x,t^*)k(x) = u_t(x,t^*). \tag{4.27}$$

If $t = t^*$ the system (4.27) becomes degenerate. For this reason, we introduce an additional condition: in the first equation of the system we assume $t \neq t^*$.

It is obvious that for small t all the systems of linear algebraic equations considered above are ill-conditioned. To solve them, it is necessary to use the regularization method (see Tikhonov, 1965a,b; Tikhonov and Arsenin, 1979; Bulygin and Danilaev, 1971).

Using natural schematic designations, we consider the ill-coditioned system of linear algebraic equations

$$ak + bk_x = c,$$

$$mk + nk_x = l, \tag{4.28}$$

which symbolically represents the systems (4.18), (4.21) and (4.22). To solve it, we apply the usual regularization algorithm (Tikhonov, 1965b). As a result, we come to solution of the system of algebraic equations

$$(a^2 + m^2 + \alpha)k + (ab + mn)k_x = ac + ml,$$

$$(ab + mn)k + (b^2 + n^2 + \alpha)k_x = bc + nl, \tag{4.29}$$

where α is the regularization parameter. Solving the system (4.29) we find

$$k = \begin{vmatrix} ac + ml & ab + mn \\ bc + nl & b^2 + n^2 + \alpha \end{vmatrix} \Bigg/ \begin{vmatrix} a^2 + m^2 + \alpha & ab + mn \\ ab + mn & b^2 + n^2 + \alpha \end{vmatrix}.$$

Assuming that the system (4.29) is a symbolic representation of the system (4.18), we obtain

$$k = \frac{\begin{vmatrix} u_x & u_t \\ \int_0^t u_x\,\mathrm{dt} & \int_0^t u_t\,\mathrm{dt} \end{vmatrix} \begin{vmatrix} u_x & u_{xx} \\ \int_0^t u_x\,\mathrm{dt} & \int_0^t u_{xx}\,\mathrm{dt} \end{vmatrix} + \alpha\left(u_{xx}u_t + \int_0^t u_{xx}\,\mathrm{dt}\int_0^t u_t\,\mathrm{dt}\right)}{\begin{vmatrix} u_x & u_{xx} \\ \int_0^t u_x\,\mathrm{dt} & \int_0^t u_{xx}\,\mathrm{dt} \end{vmatrix}^2 + \alpha\left[u_{xx}^2 + u_x^2 + \left(\int_0^t u_{xx}\,\mathrm{dt}\right)^2 + \left(\int_0^t u_x\,\mathrm{dt}\right)^2\right] + \alpha^2}. \tag{4.30}$$

As is expected, the formula (4.30) coincides with the first of formula in (4.17) when $\alpha = 0$. □

4.3. ON DERIVING THE MAIN INTEGRO-DIFFERENTIAL EQUATION

Theorem 4.6. *The form of the integro-differential equation* (3.7) *is independent of the order in which the unknown coefficient* $k(x)$ *and its derivative* $k_x(x)$ *are eliminated from the equation* (3.2).

Proof. The integro-differential equation (3.7) can be deduced by means of another method which will be used in the proof below. The structure of the derivation is as follows: we construct the system of linear algebraic equations (4.18), solve it, differentiate the solution with respect to t and equate the result to zero. Following this procedure, we solve the system (4.18) for the derivative of the unknown coefficient:

$$k_x = \begin{vmatrix} u_t & u_{xx} \\ \int_0^t u_t\,\mathrm{dt} & \int_0^t u_{xx}\,\mathrm{dt} \end{vmatrix} \Bigg/ \begin{vmatrix} u_x & u_{xx} \\ \int_0^t u_x\,\mathrm{dt} & \int_0^t u_{xx}\,\mathrm{dt} \end{vmatrix}. \tag{4.31}$$

Differentiate (4.31) with respect to t and equate the result to zero:

$$\begin{aligned} &\left(u_{tt}\int_0^t u_{xx}\,\mathrm{dt} - u_{xxt}\int_0^t u_t\,\mathrm{dt}\right)\left(u_x\int_0^t u_{xx}\,\mathrm{dt} - u_{xx}\int_0^t u_x\,\mathrm{dt}\right) \\ &- \left(u_t\int_0^t u_{xx}\,\mathrm{dt} - u_{xx}\int_0^t u_t\,\mathrm{dt}\right)\left(u_{xt}\int_0^t u_{xx}\,\mathrm{dt} - u_{xxt}\int_0^t u_x\,\mathrm{dt}\right) = 0. \end{aligned} \tag{4.32}$$

We transform (4.32) to obtain the integro-differential equation (3.7). According to the former order of operations, the system (4.18) was first written to be solved for k_x and the the condition $k_{xt} = 0$ was introduced. We change this order as follows. First write the system (4.18) to solve for k:

$$k = \begin{vmatrix} u_x & u_t \\ \int_0^t u_x \, \mathrm{dt} & \int_0^t u_t \, \mathrm{dt} \end{vmatrix} \Bigg/ \begin{vmatrix} u_x & u_{xx} \\ \int_0^t u_x \, \mathrm{dt} & \int_0^t u_{xx} \, \mathrm{dt} \end{vmatrix}.$$

The equality $k_t = 0$ leads to the equation

$$\begin{aligned} &\left(u_{xt} \int_0^t u_t \, \mathrm{dt} - u_{tt} \int_0^t u_x \, \mathrm{dt}\right)\left(u_x \int_0^t u_{xx} \, \mathrm{dt} - u_{xx} \int_0^t u_x \, \mathrm{dt}\right) \\ &- \left(u_x \int_0^t u_t \, \mathrm{dt} - u_t \int_0^t u_x \, \mathrm{dt}\right)\left(u_{xt} \int_0^t u_{xx} \, \mathrm{dt} - u_{xxt} \int_0^t u_x \, \mathrm{dt}\right) = 0. \end{aligned}$$

Transforming this equation yields

$$\begin{aligned} &(u_{xxt}u_x - u_{xx}u_{xt}) \int_0^t u_t \, \mathrm{dt} + (u_{tt}u_{xx} - u_t u_{xxt}) \int_0^t u_x \, \mathrm{dt} \\ &+ (u_{xt}u_t - u_x u_{tt}) \int_0^t u_{xx} \, \mathrm{dt} = 0. \end{aligned}$$

Thus, we obtain the same integro-differential equation (3.7), which proves the theorem. □

Theorem 4.7. *The form of the integro-differential equation* (3.7) *is independent of which of the systems* (4.18), (4.21), (4.22) *is used in the derivation of this equation.*

Proof. Let us use the system (4.21) as a basis of deriving the integro-differential equation (3.7). We follow the same order of operations: first, write the system to solve for k and introduce the equation $k_t = 0$. Then write the system to solve for k_x and introduce the equation $k_{xt} = 0$. Write all the resulting relations together in the form of the homogeneous system of linear algebraic equations for the unknowns Δ and Δ_1:

$$\Delta u_{xt} + \Delta_1 \int_0^t u_x \, \mathrm{dt} = 0,$$

$$\Delta u_{xxt} + \Delta_1 \int_0^t u_{xx} \, \mathrm{dt} = 0, \tag{4.33}$$

where

$$\Delta = \begin{vmatrix} \int_0^t u_x\,dt & \int_0^t u_{xx}\,dt & \int_0^t u_t\,dt \\ u_x & u_{xx} & u_t \\ u_{xt} & u_{xxt} & u_{tt} \end{vmatrix}, \quad \Delta_1 = \begin{vmatrix} \int_0^t u_x\,dt & \int_0^t u_{xx}\,dt & \int_0^t u_t\,dt \\ u_{xt} & u_{xxt} & u_{tt} \\ u_{xtt} & u_{xxtt} & u_{ttt} \end{vmatrix}. \tag{4.34}$$

The determinant of this system is

$$\begin{vmatrix} u_{xt} & \int_0^t u_x\,dt \\ u_{xxt} & \int_0^t u_{xx}\,dt \end{vmatrix} \neq 0.$$

Therefore, the system (4.33) has only trivial solution $\Delta = \Delta_1 = 0$. The equality $\Delta = 0$ represents the integro-differential equation (3.7). Here it is obtained on the basis of the system (4.21).

Let us prove the system (4.22) also can be used to obtain the integro-differential equation (3.7). Solve the system (4.22) for k and introduce the equation $k_t = 0$. Then solve the system (4.22) for k_x and introduce the equation $k_{xt} = 0$. Both of these equations have the same form

$$\begin{vmatrix} u_x & u_{xx} & u_t \\ u_{xt} & u_{xxt} & u_{tt} \\ u_{xtt} & u_{xxtt} & u_{ttt} \end{vmatrix} = 0. \tag{4.35}$$

Substituting $u(x,t) = \int_0^t v(x,t)\,dt$ into equation (4.35), we get

$$\begin{vmatrix} \int_0^t v_x\,dt & \int_0^t v_{xx}\,dt & \int_0^t v_t\,dt \\ v_x & v_{xx} & v_t \\ v_{xt} & v_{xxt} & v_{xx} \end{vmatrix} = 0.$$

We see that the integro-differential equation constructed for the function $v(x,t)$ coincides with equation (3.7), which proves the theorem. □

Now let us deduce the integro-differential equation (3.7) on the basis of the system (4.29). Rewrite (4.30) for calculating the coefficient k as follows:

$$k(x) = \frac{\begin{vmatrix} u_{xx}u_t + \int_0^t u_{xx}\,dt \int_0^t u_t\,dt & u_{xx}u_x + \int_0^t u_{xx}\,dt \int_0^t u_x\,dt \\ u_x u_t + \int_0^t u_x\,dt \int_0^t u_t\,dt & u_x^2 + \left(\int_0^t u_x\,dt\right)^2 + \alpha \end{vmatrix}}{\begin{vmatrix} u_{xx}^2 + \left(\int_0^t u_{xx}\,dt\right)^2 + \alpha & u_{xx}u_x + \int_0^t u_{xx}\,dt \int_0^t u_x\,dt \\ u_{xx}u_x + \int_0^t u_{xx}\,dt \int_0^t u_x\,dt & u_x^2 + \left(\int_0^t u_x\,dt\right)^2 + \alpha \end{vmatrix}}. \tag{4.36}$$

Using (4.36) and the equality $k_t = 0$, we again arrive at the integro-differential equation for the function $u(x,t)$:

$$\int_0^t u_x\,\mathrm{dt} \begin{vmatrix} \int_0^t u_x\,\mathrm{dt} & \int_0^t u_{xx}\,\mathrm{dt} & \int_0^t u_t\,\mathrm{dt} \\ u_x & u_{xx} & u_t \\ u_{xt} & u_{xxt} & u_{tt} \end{vmatrix} \begin{vmatrix} u_x & u_{xx} \\ \int_0^t u_x\,\mathrm{dt} & \int_0^t u_{xx}\,\mathrm{dt} \end{vmatrix}^2$$

$$+\alpha F(\alpha, u(x,t)) = 0, \tag{4.37}$$

where

$$\begin{aligned}
&F(\alpha, u(x,t)) \\
&= \Bigg\{ \left[u_{tt}u_{xx} + u_t u_{xxt} + u_t \int_0^t u_{xx}\,\mathrm{dt} + u_{xx}\int_0^t u_t\,\mathrm{dt} \right] \begin{vmatrix} u_x & u_{xx} \\ \int_0^t u_x\,\mathrm{dt} & \int_0^t u_{xx}\,\mathrm{dt} \end{vmatrix}^2 \\
&\quad + \left[u_{xx}^2 + u_x^2 + \left(\int_0^t u_{xx}\,\mathrm{dt}\right)^2 + \left(\int_0^t u_x\,\mathrm{dt}\right)^2 \right] \\
&\quad \times \begin{vmatrix} u_{xxt} & u_{xt} \\ \int_0^t u_{xx}\,\mathrm{dt} & \int_0^t u_x\,\mathrm{dt} \end{vmatrix} \begin{vmatrix} u_t & u_x \\ \int_0^t u_t\,\mathrm{dt} & \int_0^t u_x\,\mathrm{dt} \end{vmatrix} \\
&\quad + \left[u_{xx}^2 + u_x^2 + \left(\int_0^t u_{xx}\,\mathrm{dt}\right)^2 + \left(\int_0^t u_x\,\mathrm{dt}\right)^2 \right] \\
&\quad \times \begin{vmatrix} u_{xx} & u_x \\ \int_0^t u_{xx}\,\mathrm{dt} & \int_0^t u_x\,\mathrm{dt} \end{vmatrix} \begin{vmatrix} u_{tt} & u_{xt} \\ \int_0^t u_t\,\mathrm{dt} & \int_0^t u_x\,\mathrm{dt} \end{vmatrix} \\
&\quad - 2\left[u_{xx}u_{xxt} + u_x u_{xt} + u_{xx}\int_0^t u_{xx}\,\mathrm{dt} + u_x \int_0^t u_x\,\mathrm{dt} \right] \\
&\qquad \times \begin{vmatrix} u_{xx} & u_x \\ \int_0^t u_{xx}\,\mathrm{dt} & \int_0^t u_x\,\mathrm{dt} \end{vmatrix} \begin{vmatrix} u_t & u_x \\ \int_0^t u_t\,\mathrm{dt} & \int_0^t u_x\,\mathrm{dt} \end{vmatrix} \\
&\quad - 2\left[u_t u_{xx} + \int_0^t u_t\,\mathrm{dt} \int_0^t u_{xx}\,\mathrm{dt} \right] \\
&\qquad \times \begin{vmatrix} u_x & u_{xx} \\ \int_0^t u_x\,\mathrm{dt} & \int_0^t u_{xx}\,\mathrm{dt} \end{vmatrix} \begin{vmatrix} u_{xt} & u_{xxt} \\ \int_0^t u_t\,\mathrm{dt} & \int_0^t u_{xx}\,\mathrm{dt} \end{vmatrix} \Bigg\} \\
&\quad + \alpha \Bigg\{ \left[u_{tt}u_{xx} + u_t u_{xxt} + u_t \int_0^t u_{xx}\,\mathrm{dt} + u_{xx}\int_0^t u_t\,\mathrm{dt} \right] \\
&\qquad \times \left[u_{xx}^2 + u_x^2 + \left(\int_0^t u_{xx}\,\mathrm{dt}\right)^2 + \left(\int_0^t u_x\,\mathrm{dt}\right)^2 \right]
\end{aligned}$$

$$
\begin{aligned}
&+\begin{vmatrix} u_{xxt} & u_{xt} \\ \int_0^t u_{xx}\,\mathrm{d}t & \int_0^t u_x\,\mathrm{d}t \end{vmatrix} \begin{vmatrix} u_t & u_x \\ \int_0^t u_t\,\mathrm{d}t & \int_0^t u_x\,\mathrm{d}t \end{vmatrix} \\
&+\begin{vmatrix} u_{xx} & u_x \\ \int_0^t u_{xx}\,\mathrm{d}t & \int_0^t u_x\,\mathrm{d}t \end{vmatrix} \begin{vmatrix} u_{tt} & u_{xt} \\ \int_0^t u_t\,\mathrm{d}t & \int_0^t u_x\,\mathrm{d}t \end{vmatrix} \\
&-2\Big[u_{xx}u_{xxt} + u_x u_{xt} + u_{xx}\int_0^t u_{xx}\,\mathrm{d}t + u_x \int_0^t u_x\,\mathrm{d}t\Big] \\
&\qquad \times \Big[u_t u_{xx} + \int_0^t u_t\,\mathrm{d}t \int_0^t u_{xx}\,\mathrm{d}t\Big]\Big\} \\
+\,\alpha^2 &\Big[u_{tt}u_{xx} + u_t u_{xxt} + u_t \int_0^t u_{xx}\,\mathrm{d}t + u_{xx}\int_0^t u_t\,\mathrm{d}t\Big].
\end{aligned}
$$

Assume that

$$
\begin{vmatrix} u_x & u_{xx} \\ \int_0^t u_x\,\mathrm{d}t & \int_0^t u_{xx}\,\mathrm{d}t \end{vmatrix} \neq 0
$$

as before, and, furthermore, assume $\int_0^t u_x\,\mathrm{d}t \neq 0$. Then, setting $\alpha = 0$ for the regularization parameter in the obtained integro-differential equation, we arrive at the original integro-differential equation (3.7).

Obviously, the form of equation (4.37) is more complicated as compared to equation (3.7). Solving the former by the quasi-inversion method, we obtain even more complicated equation. However, this is not always necessary. After solving the integro-differential equation (3.7) deduced above, we can use the formula (4.36) to determine the sought function, i.e., the coefficient $k(x)$.

Chapter 5.

Generalizations of the developed algorithm for solving coefficient inversion problems

In this chapter we point out the possible generalizations and ways of using the algorithm of solving the coefficient inversion problem for equation (3.2) which was developed in Chapter 3 For for solving other coefficient inverse problems. In particular, we consider the problem of finding the coefficient $k(r)$ in the equation

$$\frac{1}{r}[rk(r)u_r]_r = u_t$$

and solve the coefficient inversion problem for the parabolic equation of more general form

$$a(x)u_{xx} + b(x)u_x = u_t + f(x,t), \tag{5.1}$$

with two unknown coefficients $a(x)$ and $b(x)$.

These generalizations and their extension to equations of another form, in particular, to the equation

$$[k(x)u_x]_x = u_t + f(x,t), \tag{5.2}$$

will be also considered further in Chapter 6 in connection with applications of these problems.

So, we consider the coefficient inverse problem for the above equation of the heat conduction equation type in the polar coordinate system, with the unknown coefficient depending only on the radius. The problem under consideration is applied in the study of model problems of radial filtration to a single central well with the usual additional condition: the water permeability coefficient, or filtration factor, is assumed to depend only on the radius.

Consider the equation

$$[k(r)u_r]_r + \frac{1}{r}k(r)u_r = u_t. \tag{5.3}$$

Rewrite (5.3) in the form

$$ku_{rr} + \left(k_r + \frac{k}{r}\right)u_r = u_t \tag{5.4}$$

and formulate a coefficient inverse problem for equation (5.4): determine the vector function $\{k(r), u(r,t)\}$ from the conditions

$$ku_{rr} + k_1 u_r = u_t, \qquad r_0 < r < R, \quad -\varepsilon < t \le t_1, \quad k_1(r) = k_r + \frac{k}{r},$$

$$\begin{aligned} u(r_0,t) &= g_0(t), \qquad u_r(r_0,t) = h_0(t), \qquad -\varepsilon < t \le t_1,\\ u(R,t) &= g_1(t), \qquad u_r(R,t) = h_1(t), \qquad -\varepsilon < t \le t_1,\\ u(r,0) &= \varphi(r), \qquad r_0 \le r \le R,\\ g_0(0) &= \varphi(r_0), \qquad g_1(0) = \varphi(R),\\ h_0(0) &= \varphi_r(r_0), \qquad h_1(0) = \varphi_r(R). \end{aligned} \tag{5.5}$$

Transform equation (5.4) to the form that does not include the coefficient $k(r)$, as in Section 3.1 . The problem is reduced, as before, to solution of the non-linear integro-differential equation with the corresponding set of additional conditions:

$$Av_{rr} + Bv_r = Cv_t, \qquad r_0 < r < R, \qquad -\varepsilon < t \le t_1,$$

$$\begin{aligned} v(r_0,t) &= \mu_0(t), \qquad v_r(r_0,t) = \eta_0(t), \qquad -\varepsilon < t \le t_1,\\ v(R,t) &= \mu_1(t), \qquad v_r(R,t) = \eta_1(t), \qquad -\varepsilon < t \le t_1, \end{aligned}$$

here $v(r,0)$ is defined arbitrarily. The problem (5.5) differs from the problem considered in Section 3 only by the designation of the independent variable.

Therefore, the designations introduced when transferring to the integro-differential equation are natural and analogous to those in Section 3.2. Further solution of the problem is carried out as in Chapter 3.

The coefficient inverse problem for equation (5.1) is formulated as follows: Find the vector function $\{a(x), b(x), u(x,t)\}$ from the conditions

$$
\begin{gathered}
a(x)u_{xx} + b(x)u_x = u_t + f(x,t), \qquad x_0 < x < x_1, \quad -\varepsilon < t \le t_1, \\
u(x_0,t) = g_0(t), \qquad u_x(x_0,t) = h_0(t), \qquad -\varepsilon < t \le t_1, \\
u(x_1,t) = g_1(t), \qquad u_x(x_1,t) = h_1(t), \qquad -\varepsilon < t \le t_1, \\
u(x,0) = \varphi(x), \qquad x_0 \le r \le x_1, \\
g_0(0) = \varphi(x_0), \qquad g_1(0) = \varphi(x_1), \\
h_0(0) = \varphi_x(x_0), \qquad h_1(0) = \varphi_x(x_1). \qquad (5.6)
\end{gathered}
$$

The algorithm for solving coefficient inversion problem developed in Chapter 3 does not use the differential dependence between the coefficient $k(x)$ in equation (3.2) and its derivative $k_x(x)$. It allows determining them as independent coefficients. Therefore, this algorithm can be applied for solving the problem (5.6).

Chapter 6.

On applications of coefficient inverse problems in underground fluid dynamics

6.1. DETERMINATION OF FILTRATION PARAMETERS OF EXPLOITED NON-HOMOGENEOUS OIL STRATA

One of the important parameters of the operating state of an exploited oil deposit is pressure distribution in a stratum (or strata that make up the deposit), the so-called field of the stratum pressure. Estimating pressure distribution in a stratum makes it possible to plan the operations on this stratum in order to enhance and control the oil production process.

By the direct problem of subsurface fluid dynamics with nonstationary filtration of compressible fluid we mean the problem of determining the stratum pressure on condition that the fields of filtration parameters that characterize filtration properties of the porous medium are known and the function of distributed source-sink is defined, i.e., the ratings of wells are known.

By the inverse problem of subsurface fluid dynamics in the case of nonstationary filtration of compressible fluid we mean either of the following problems:

— the problem of determining the vector function with components represented by the function of the stratum pressure and one of the filtration parameters, for example, the water permeability coefficient as a complex parameter; the function of distributed source-sink is assumed to be known;

—the problem of determining the vector function with components represented by the function of the stratum pressure and the density function of distributed source-sink; the fields of filtration parameters, for example, the field of the water permeability coefficient, are assumed to be known;

— the problem of determining the vector function with components represented by the function of the stratum pressure, the unknown coefficient characterizing filtration properties of the porous medium, for example, the water permeability coefficient, and the function of distributed source-sink.

All the listed problems are coefficient inverse problems, for example, for the following parabolic equation:

$$\frac{\partial}{\partial x}\left(\sigma(x)\,\frac{\partial p}{\partial x}\right) = a\,\frac{\partial p}{\partial t} + f(x,t), \tag{6.1}$$

which describes the process of non-stationary single-phase one-dimensional filtration of compressible fluid in a heterogeneous porous medium, where $p(x,t)$ is the function of the stratum pressure, $f(x,t)$ is the function of distributed source-sink, $\sigma(x)$ is the water permeability coefficient, $\sigma = kh/\mu$, k is the porous permeability index, h is the thickness of the oil stratum, μ is the coefficient of dynamic viscosity of filtered fluid. This coefficient is defined in a more complicated way in the cases of the multi-phase filtration and filtration of anomalous fluids.

So, for equation (6.1) these problems are reduced to finding the vector functions $\{\sigma(x), p(x,t)\}$, $\{f(x,t), p(x,t)\}$, and $\{\sigma(x), f(x,t), p(x,t)\}$, correspondingly. The problems are conditionally well-posed and require special methods of solution. Solving these problems makes it possible to "pre-estimate" the working condition of an oil stratum and make propositions on how to operate it.

We now turn to the first of the listed problems. The water permeability coefficient has an important property: as a rule, it varies slowly with time in the single-phase filtration region so that this dependence can be neglected for a rather large time interval. We shall assume in what follows that $\sigma = \sigma(x)$. This makes it possible to use the uniqueness theorems formulated and proved in Klibanov (1984a, 1984b, 1986) and Klibanov and Danilaev (1990) when studying the coefficient inverse problem for equation (6.1) with the sought vector function $\{\sigma(x), p(x,t)\}$.

We now consider non-stationary filtration of compressible fluid in a heterogeneous porous medium which is governed by different laws.

Analyze the second of problems, in which $\{f(x,t), p(x,t)\}$ is the sought function. The uniqueness of a solution to this problem is studied in Savateev (1995), where the sought function $f(x,t)$ was assumed to have a special form. This question was also studied in Lavrent'ev, Reznitskaya, and Yakhno (1982).

We estimate the role of the approach that we propose as compared with other methods of determining the fields of filtration parameters, pointing out the most important ones.

In Bulygin (1958), two methods are proposed for calculating water permeability on the basis of the data of engineering documentation of an exploited oil deposit. The first one is formulated as the Cauchy problem for a partial differential equation of the first order written for the unknown coefficient. The equation that describes the filtration process can be reduced to this problem after the function of the stratum pressure is defined. This approach was developed by G. V. Golubev in Golubev, Danilaev, and Tumashev (1978), Golubev (1992), Golubev and Danilaev (1981, 1983, 1987, 1990, 1991a, 1991b, 1992, 1996a, 1996b). He developed the numerical methods for solving this problem in the above-mentioned setting and studied it.

The second approach (we call it the statistical treatment of the observation results) implies most extensive use of the engineering data at the researcher's disposal. First, the function of the stratum pressure is defined in the equation describing the filtration process. Then the transformed equation is written as many times as the number of available isobaric maps. As a result, we get an overdetermined and, as a rule, ill-conditioned system of linear algebraic equations for the water permeability coefficient. This approach was developed in Bulygin and Danilaev (1971), Danilaev (1978)and Golubev, Danilaev, and Tumashev (1978).

In Butkovskii (1975) and in Bellman, Gluss, and Roth (1965), Bellman, Kagiwada, Kalaba,and Ueno (1965), Cahen (1963), Douglas and Jones (1962), Kagivada (1964) the problem of determining the fields of the unknown coefficients is formulated as an optimization problem. Further development of the mathematical methods for determining the fields of the unknown coefficient was connected with the use of regularization methods. In this process, the underlying principles of the conditional well-posedness of the problems of determining the unknown coefficients for parabolic equations were taken into account. These investigations were carried out by Khairullin (1986, 1988), Chavent (1970, 1971), Chavent, Dupuy, and Lemonnier (1975), Kravaris and Seinfeld (1982, 1985, 1986). For example, in Kravaris and Sienfeld (1985, 1986) the problem of determining the unknown coefficients in equation (6.1) is formulated as a problem of parametric identification and solved by the regularization method.

The distinctive feature of the proposed approach is that the results of the proofs of uniqueness of solutions to the corresponding coefficient inverse problems are used in combination with special methods of solution of conditionally well-posed problems.

Consider how the developed algorithm is applied in determining the field of water permeability coefficient for one-dimensional non-stationary filtration of compressible fluid following Darcy's law in a heterogeneous porous medium. In the considering case the filtration process is described by equation (6.1) in the domain $(x_0, x_1) \times (0, t_1]$. The necessary set of initial and boundary conditions is assumed to be given:

$$p(x_0, t) = g_0(t), \qquad \frac{\partial p}{\partial x}(x_0, t) = h_0(t), \qquad t \in (-\varepsilon, t_1],$$

$$p(x_1, t) = g_1(t), \qquad \frac{\partial p}{\partial x}(x_1, t) = h_1(t), \qquad t \in (-\varepsilon, t_1],$$

$$p(x, 0) = \varphi(x), \qquad x \in [x_0, x_1].$$

Here $g_0(t)$, $h_0(t)$, $g_1(t)$, $h_1(t)$, $\varphi(x)$ are given functions satisfying the usual conditions. Applying the algorithm presented in Chapter 3, we formulate and solve the coefficient inverse problem for equation (6.1) and find the vector function $\{\sigma(x), p(x, t)\}$.

As was noted before, the quasi-inversion method can be used for solving the problem of continuation of a solution of a parabolic equation, to which the solution of the coefficient inverse problem is reduced, only for small time half-intervals. Assume that the solution of the initial-boundary value problem for equation (6.1), i.e. the function $p(x, t)$, varies slowly with time. Taking into account the given initial condition $p(x, 0)$, in this case it is reasonable to speak only of the function $\sigma(x)$ determined as a component of the sought vector function. The developed algorithm allows determining the unique function $\sigma(x)$ which is stable for small variations of initial data.

On a small time interval, the functions $g_0(t)$, $h_0(t)$, $g_1(t)$, $h_1(t)$ can be assumed to be constant, equal to their values at $t = 0$. As a rule, in practice, engineering data allows defining these values at a number of discrete instants of time. The number of these points can be large. In this case a question arises concerning the use of all the available information in order to to provide high accuracy for the obtained solution. A possible answer to this question will be given below.

Using the data given for a number of discrete instants of time, for each of them we formulate a coefficient inverse problem. Solving these problems, we get the corresponding distribution for the water permeability coefficient. Then we use statistical methods to find the most probable distribution of the water permeability coefficient as an estimate of the mathematical expectation of distributions determined in the process of solution of these coefficient inverse problems. Such a distribution will use the whole amount of the given data.

6.2. DETERMINING THE FILTRATION PARAMETERS IN THE CASE OF NON-LINEAR FILTRATION

Consider the non-linear filtration with initial pressure gradient. Such filtration is also called anomalous fluid filtration or filtration that does not follow linear Darcy's law. There are different laws that govern the filtration $\vec{v}$ in the theory of filtration of non-Newtonian fluids. We use the following law:

$$\vec{v} = -\frac{k}{\mu}\Big(1 - \frac{\beta}{|\nabla p|}\Big)\nabla p. \tag{6.2}$$

As before, here k is the permeability index of the porous medium, μ is the dynamic viscosity of the filtered fluid, $\beta = \beta(x)$ is the given variable rate of shear. In this case, the main filtration equation describing the stratum pressure distribution for one-dimensional flow along the abscissa axis is written as follows:

$$\frac{\partial}{\partial x}\Big[k\Big(\frac{\partial p}{\partial x} - \beta \operatorname{sign}\Big(\frac{\partial p}{\partial x}\Big)\Big)\Big] = \frac{1}{a}\frac{\partial p}{\partial t}. \tag{6.3}$$

We shall assume $a = 1$.

Consider the problem of finding the permeability index k entering the principle divergent part of equation (6.3) as one of its coefficient. We assume that $k = k(x)$, i.e. k is a function of x only and does not depend on t. Following the developed theory, we formulate the problem: Determine the vector function $\{k(x), p(x,t)\}$ from the conditions

$$\frac{\partial}{\partial x}\Big[k\Big(\frac{\partial p}{\partial x} - \beta \operatorname{sign}\Big(\frac{\partial p}{\partial x}\Big)\Big)\Big] = \frac{\partial p}{\partial t}, \qquad x \in (x_0, x_1), \quad t \in (-\varepsilon, t_1],$$

$$p(x_0,t) = g_0(t), \qquad \frac{\partial p}{\partial x}(x_0,t) = h_0(t), \qquad t \in (-\varepsilon, t_1],$$

$$p(x_1,t) = g_1(t), \qquad \frac{\partial p}{\partial x}(x_1,t) = h_1(t), \qquad t \in (-\varepsilon, t_1],$$

$$p(x,0) = \varphi(x), \qquad x \in [x_0, x_1],$$

where $g_0(0) = \varphi(x_0)$, $g_1(0) = \varphi(x_1)$, $h_0(0) = \varphi'_x(x_0)$ and $h_1(0) = \varphi'_x(x_1)$. Transform equation (6.3) to eliminate $k(x)$. Before this, we introduce a new unknown function instead of $p(x,t)$:

$$u(x,t) = p(x,t) - \int_{x_0}^{x} \beta(x) \operatorname{sign}\left(\frac{\partial p}{\partial x}\right) dx. \tag{6.4}$$

With the function $u(x,t)$, equation (6.3) takes the form

$$[k(x)u_x]_x = u_t.$$

It is obvious that without source-sink in the domain $x \in (x_0, x_1)$ the sign $\operatorname{sign}(\partial p/\partial x)$ remains unchanged throughout the entire domain of filtration (solution). It is completely determined by defining the value of the derivative $\partial p/\partial x$ on the boundary of the domain.

Suppose that there are no stagnation zones inside the filtration region. This allows us, instead of defining the necessary set of boundary conditions for the function $p(x,t)$, to define the corresponding boundary conditions for the new unknown function $u(x,t)$:

$$u(x_0,t) = g_0(t), \qquad \frac{\partial u}{\partial x}(x_0,t) = h_0(t),$$

$$u(x_1,t) = \tilde{g}_1(t) = g_1(t) - \operatorname{sign}\left(\frac{\partial p}{\partial x}(x_1,t)\right) \int_{x_0}^{x_1} \beta(x)\, dx,$$

$$\frac{\partial u}{\partial x}(x_1,t) = \tilde{h}_1(t) = h_1(t) - \operatorname{sign}\left(\frac{\partial p}{\partial x}(x_1,t)\right) \int_{x_0}^{x_1} \beta(x)\, dx,$$

$$u(x,0) = \tilde{\varphi}(x) = \tilde{\varphi}(x) - \operatorname{sign}\left(\frac{\partial p}{\partial x}(x_1,t)\right) \int_{x_0}^{x_1} \beta(x)\, dx, \tag{6.5}$$

Further solution of the problem using the quasi-inversion method is reduced to the procedures described above.

The function of the stratum pressure at the bore points has logarithmic singularities. In order to solve the problem of determining the stratum pressure, these singularities should be eliminated, i. e., the function should be made continuous. For this purpose, subsurface fluid dynamics involves the well-known technique of "spreading" the source-sinks by introducing the function $f(x,t)$ of the density of distributed source-sinks in the "material balance" equation (6.3). There are the different empirical formulas to introduce this function (Danilaev, 1978). We shall not list these formulas and techniques here, since it is not the subject of our study. We write equation (6.3) with introduced function of the density of distributed extraction, as it is also called:

$$\frac{\partial}{\partial x}\left[k\left(\frac{\partial \tilde{p}}{\partial x} - \beta \operatorname{sign}\left(\frac{\partial \tilde{p}}{\partial x}\right)\right)\right] = \frac{\partial \tilde{p}}{\partial t} + f(x,t). \tag{6.6}$$

Here the sought function $\tilde{p}(x,t)$ has no logarithmic singularities at the bore points. As before, we shall assume that $\operatorname{sign}(\partial\tilde{p}/\partial x)$ remains unchanged in the entire domain of solution, i. e., the sign of the derivative $\partial\tilde{p}/\partial x$ does not change in the region of filtration (solution domain).

The statement of the problem of determining the unknown coefficient $k(x)$ for equation (6.6) remains the same as in the previous case. After the substitution of the form (6.4), i. e., introducing the new unknown function $\tilde{u}(x,t)$, we transform equation (6.6) to the form

$$\frac{\partial}{\partial x}\left(k\frac{\partial \tilde{u}}{\partial x}\right) = \frac{\partial \tilde{u}}{\partial t} + f(x,t).$$

Initial and boundary conditions have the same form (6.5).

The algorithm of solution of this problem remains the same. As in Chapter 3, we introduce the new unknown function $v = u_t/u$ and, eliminating the unknown coefficient from the equation, transform it to the non-linear integro-differential equation of the forms slightly different from that of equation (3.7):

$$A_1 v_{xx} + B_1 v_x = C v_t + (Mv + N),$$

where

$$A_1 = EJ_0 + \left(v + \frac{1}{1-E}\right)J_1 = A + \frac{J_1}{1-E}, \qquad B_1 = B - f\,\frac{J_3}{1-E},$$

$$C = J_1\left[J_0^2 - \left(\frac{\partial \ln\tilde{\varphi}}{\partial x}\right)^2 + J_4\right] - J_0J_3, \quad M = -f\,\frac{C + 2J_1J_0}{1-E}, \quad N = f_t\,\frac{C}{1-E}.$$

Here we used the notation of Chapter 3.

Consider another way of using the developed algorithm of solving coefficient inverse problem for anomalous systems. We solve the formulated problem without introducing the new unknown function $u(x,t)$, making transformations similar to those made for the corresponding equation in the case of linear filtration, as in Chapter 3. Determine the vector function $\{k(x), p(x,t)\}$ from the conditions

$$\frac{\partial}{\partial x}\left[k\left(\frac{\partial p}{\partial x} - \beta \operatorname{sign}\left(\frac{\partial p}{\partial x}\right)\right)\right] = \frac{\partial p}{\partial t}, \qquad x \in (x_0, x_1), \qquad t \in (-\varepsilon, t_1],$$

$$p(x_0,t)=g_0(t),\qquad \frac{\partial p}{\partial x}(x_0,t)=h_0(t),\qquad t\in(-\varepsilon,t_1],$$

$$p(x_1,t)=g_1(t),\qquad \frac{\partial p}{\partial x}(x_1,t)=h_1(t),\qquad t\in(-\varepsilon,t_1],$$

$$p(x,0)=\varphi(x),\qquad x\in[x_0,x_1],$$

where $g_0(t)$, $g_1(t)$, $h_0(t)$, $h_1(t)$ are given functions and the following concordance conditions hold:

$$g_0(0)=\varphi(x_0),\qquad g_1(0)=\varphi(x_1),\qquad h_0(0)=\varphi'_x(x_0),\qquad h_1(0)=\varphi'_x(x_1)$$

As in Section 3.1, we transform equation (6.6) to eliminate the unknown coefficient $k(x)$. For this purpose, we rewrite equation (6.6) in the form

$$k(p'_{xx}-\beta'_x\operatorname{sign}(p'_x))+k'_x(p'_x-\beta\operatorname{sign}(p'_x))=p'_t$$

and integrate with respect to time from 0 to the variable upper limit t:

$$k\int_0^t(p'_{xx}-\beta'_x\operatorname{sign}(p'_x))\,\mathrm{d}t+k'_x\int_0^t(p'_x-\beta\operatorname{sign}(p'_x))\,\mathrm{d}t=p-\varphi.$$

According to the procedure in Section 3.1, we write the obtained expression with $k(x)$ on the left-hand side and differentiate the result with respect to t observing that $k_t=0$. We have

$$\begin{aligned}p_t\int_0^t&(p_{xx}-\beta_x\operatorname{sign}(p_x))\,\mathrm{d}t-(p-\varphi)(p_{xx}-\beta_x\operatorname{sign}(p_x))\\&=k_x\Big[(p_{xx}-\beta_x\operatorname{sign}(p_x))\int_0^t(p_{xx}-\beta_x\operatorname{sign}(p_x))\,\mathrm{d}t\\&\qquad-(p_{xx}-\beta_x\operatorname{sign}(p_x))\int_0^t(p_x-\beta\operatorname{sign}(p_x))\,\mathrm{d}t\Big].\end{aligned}$$

Solve the obtained equality for $k_x(x)$:

$$\begin{aligned}k_x=&\Big[p_t\int_0^t(p_{xx}-\beta_x\operatorname{sign}(p_x))\,\mathrm{d}t-(p-\varphi)(p_{xx}-\beta_x\operatorname{sign}(p_x))\Big]\\&\times\Big[(p_{xx}-\beta_x\operatorname{sign}(p_x))\int_0^t(p_{xx}-\beta_x\operatorname{sign}(p_x))\,\mathrm{d}t\\&\qquad-(p_{xx}-\beta_x\operatorname{sign}(p_x))\int_0^t(p_x-\beta\operatorname{sign}(p_x))\,\mathrm{d}t\Big]^{-1}.\end{aligned}$$

Differentiate the resulting equality with respect to t once more. Using the condition $k_{xt} = 0$, after transformations we arrive at the integro-differential equation

$$
\begin{aligned}
&(p-\varphi)(p_{xt}p_{xx} - p_{xt}\beta_x \operatorname{sign}(p_x)) - (p_x p_{xxt} - p_{xxt}\beta \operatorname{sign}(p_x)) \\
&+ (p_t p_{xxt} - p_{tt}p_{xx} + p_{tt}\beta_x \operatorname{sign}(p_x)) \int_0^t (p_x - \beta \operatorname{sign}(p_x))\,dt \\
&+ (p_{tt}p_x - p_t p_{xt} - p_{tt}\beta \operatorname{sign}(p_x)) \int_0^t (p_{xx} - \beta_x \operatorname{sign}(p_x))\,dt = 0. \quad (6.7)
\end{aligned}
$$

One can verify that equation (3.6), corresponds to the case of nonlinear filtration, is a special case of the foregoing equation (6.7) with $\beta = 0$ (and, consequently, $\beta_x = 0$). Equation (6.7) can be written in a simpler form, taking into account that

$$
\int_0^t \beta \operatorname{sign}(p_x)\,dt = \beta t \operatorname{sign}(p_x), \qquad \int_0^t \beta_x \operatorname{sign}(p_x)\,dt = \beta_x t \operatorname{sign}(p_x)
$$

It takes the form

$$
\begin{aligned}
&(p-\varphi)(p_{xt}p_{xx} - p_{xt}\beta_x \operatorname{sign}(p_x)) - (p_x p_{xxt} - p_{xxt}\beta \operatorname{sign}(p_x)) \\
&+ (p_t p_{xxt} - p_{tt}p_{xx} + p_{tt}\beta_x \operatorname{sign}(p_x))\Big(\int_0^t p_x\,dt - \beta_x t \operatorname{sign}(p_x)\Big) \\
&+ (p_{tt}p_x - p_t p_{xt} - p_{tt}\beta \operatorname{sign}(p_x))\Big(\int_0^t p_{xx}\,dt - \beta_x t \operatorname{sign}(p_x)\Big) = 0. \quad (6.8)
\end{aligned}
$$

Thus, we obtained the non-linear integro-differential equation for the function $p(x,t)$ without unknown coefficient. As before, we transform it by introducing the new unknown function $v(x,t) = p_t(x,t)/p(x,t)$. After the substitution, we get the following expressions:

$$
p(x,t) = \varphi(x)\exp\Big(\int_0^t v(x,t)\,dt\Big), \qquad \frac{p_x}{p} = \frac{\varphi_x}{\varphi} + \int_0^t v_x\,dt,
$$

$$
\frac{p_{xx}}{p} = \Big(\frac{\varphi_x}{\varphi} + \int_0^t v_x\,dt\Big)^2 - \Big(\frac{\varphi_x}{\varphi}\Big)^2 + \Big(\frac{\varphi_{xx}}{\varphi} + \int_0^t v_{xx}\,dt\Big),
$$

$$
\frac{p_{xt}}{p} = \Big(\frac{\varphi_x}{\varphi} + \int_0^t v_x\,dt\Big)v + v_x, \qquad \frac{p_t}{p} + v_t + v^2,
$$

$$
\begin{aligned}
\frac{p_{xxt}}{p} = v\Big[&\Big(\frac{\varphi_x}{\varphi} + \int_0^t v_x\,dt\Big)^2 - \Big(\frac{\varphi_x}{\varphi}\Big)^2 + \Big(\frac{\varphi_{xx}}{\varphi} + \int_0^t v_{xx}\,dt\Big)\Big] \\
&+ 2v_x\Big(\frac{\varphi_x}{\varphi} + \int_0^t v_x\,dt\Big) + v_{xx}.
\end{aligned}
$$

We use the notation introduced before:

$$E = 1 - \exp\Big(\int_0^t v(x,t)\,\mathrm{d}t\Big),\ J_0 = \int_0^t v_x\,\mathrm{d}t + (\ln\varphi)_x,\ J_1 = \int_0^t (1-E)J_0\,\mathrm{d}t,$$

$$J_4 = \int_0^t v_{xx}\,\mathrm{d}t + \frac{\varphi_x}{\varphi}, \qquad J_2 = J_0^2 - \Big(\frac{\varphi_x}{\varphi}\Big)^2 + J_4, \qquad J_3 = \int_0^t (1-E)J_2\,\mathrm{d}t.$$

With this notation we have

$$p(x,t) = \varphi(x)(1-E), \qquad \frac{p_x}{p} = J_0, \qquad \frac{p_{xx}}{p} = J_2, \qquad \frac{p_{xt}}{p} = J_0 v + v_x,$$

$$\frac{p_{xxt}}{p} = J_2 v + 2J_0 v_x + v_{xx}, \qquad \int_0^t p_x\,\mathrm{d}t = \varphi J_1,$$

$$\int_0^t p_{xx}\,\mathrm{d}t = \varphi J_3, \qquad \operatorname{sign}(p_x) = \operatorname{sign}(J_0\varphi).$$

Dividing both sides of equation (6.8) by p^2, we rewrite it using the new unknown function $v(x,t)$:

$$A_1 v_{xx} + B_1 v_x = C_1 v_t + Mv + Nv^2. \tag{6.9}$$

Here we use the following notation:

$$A_1 = A\varphi - \Big(\frac{E}{1-E} + vt)\beta \operatorname{sign}(\varphi J_0),$$

$$B_1 = B\varphi - \Big(\frac{E}{1-E} + vt\Big)(\beta_x - 2\beta J_0)\operatorname{sign}(\varphi J_0),$$

$$C_1 = C\varphi - \Big(\frac{J_3}{1-E} + J_2 t\Big)\operatorname{sign}(\varphi J_0)\beta - \Big(\frac{J_1}{1-E} + J_0 t\Big)\beta_x \operatorname{sign}(\varphi J_0),$$

$$M = \frac{E}{1-E}(\beta J_2 - \beta_x J_0)\operatorname{sign}(\varphi J_0),$$

$$N = \frac{E}{1-E}(\beta J_3 - \beta_x J_1)\operatorname{sign}(\varphi J_0).$$

The coefficients A, B and C are the same as in equation (3.7) (Section 3.1). It is easy to verify that for $\beta = 0$ equation (6.9) can be transformed to equation (3.7).

Now consider coefficient inverse problems for anomalous systems with a special form of the filtration law. We mean the filtration described by equation (6.3) with $\beta = \alpha/\sqrt{k}$, where $\alpha = 1$ and α is a given constant. It was assumed above that $\beta(x)$ is a given function. We formulate the coefficient inverse problem for equation (6.3), its setting remaining the same.

Rewrite equation (6.3) as follows

$$k\frac{\partial^2 p}{\partial x^2} + \frac{\partial k}{\partial x}\frac{\partial p}{\partial x} - \frac{\alpha}{2\sqrt{k}}\frac{\partial k}{\partial x}\,\mathrm{sign}\left(\frac{\partial p}{\partial x}\right) - \frac{\alpha}{\sqrt{k}}\frac{\partial}{\partial x}\left(\mathrm{sign}\left(\frac{\partial p}{\partial x}\right)\right) = \frac{\partial p}{\partial t}.$$

Assuming again that there are no source-sink in the filtration region, we conclude that

$$\frac{\partial}{\partial x}\left(\mathrm{sign}\left(\frac{\partial p}{\partial x}\right)\right) = 0.$$

We shall assume that the derivative $\partial p/\partial x$ does not change its sign on the interval $(0, t_1]$. This means that the quantity

$$\frac{\alpha}{2\sqrt{k}}\frac{\partial k}{\partial x}\,\mathrm{sign}\left(\frac{\partial p}{\partial x}\right) = f(x)$$

does not depend on time. Thus, the problem is formulated as follows: Determine the vector function $\{k(x), p(x,t)\}$ satisfying the equation

$$k\frac{\partial^2 p}{\partial x^2} + \frac{\partial k}{\partial x}\frac{\partial p}{\partial x} - \frac{\partial p}{\partial t} = f(x), \tag{6.10}$$

with the same additional conditions (6.5).

Eliminate the coefficient k from equation (6.10). To this end, first, differentiate it with respect to t, taking into account that $[f(x)]_t = 0$. We obtain

$$kp_{xxt} + k_x p_{xt} - p_{tt} = 0. \tag{6.11}$$

Integrate equation (6.11) twice with respect to t from 0 to t so that the integrals have variable upper limits.

$$k\int_0^t\left(\int_0^t p_{xxt}\,\mathrm{d}t\right)\mathrm{d}t + k_x\int_0^t\left(\int_0^t p_{xt}\,\mathrm{d}t\right)\mathrm{d}t - \int_0^t\left(\int_0^t p_{tt}\,\mathrm{d}t\right)\mathrm{d}t = 0.$$

Write this equation with the derivative k_x on the left-hand side and differentiate it with respect to t observing that $k_{xt} = 0$:

$$k_x = \left[\int_0^t\left(\int_0^t p_{tt}\,\mathrm{d}t\right)\mathrm{d}t - k\int_0^t\left(\int_0^t p_{xxt}\,\mathrm{d}t\right)\mathrm{d}t\right]\Big/\int_0^t\left(\int_0^t p_{xt}\,\mathrm{d}t\right)\mathrm{d}t,$$

$$\left(\int_0^t p_{tt}\,\mathrm{d}t - k\int_0^t p_{xxt}\,\mathrm{d}t\right)\int_0^t\left(\int_0^t p_{xt}\,\mathrm{d}t\right)\mathrm{d}t$$
$$-\left[\int_0^t\left(\int_0^t p_{tt}\,\mathrm{d}t\right)\mathrm{d}t - k\int_0^t\left(\int_0^t p_{xxt}\,\mathrm{d}t\right)\mathrm{d}t\right]\int_0^t p_{xt}\,\mathrm{d}t = 0.$$

Solve the resulting equation for the coefficient k:

$$
\begin{aligned}
k(x) = &\left[\int_0^t p_{xt}\,\mathrm{dt}\int_0^t\left(\int_0^t p_{tt}\,\mathrm{dt}\right)\mathrm{dt} - \int_0^t\left(\int_0^t p_{xt}\,\mathrm{dt}\right)\mathrm{dt}\int_0^t p_{tt}\,\mathrm{dt}\right]\\
&\times\left[\int_0^t p_{xt}\,\mathrm{dt}\int_0^t\left(\int_0^t p_{xxt}\,\mathrm{dt}\right)\mathrm{dt} - \int_0^t\left(\int_0^t p_{xt}\,\mathrm{dt}\right)\mathrm{dt}\int_0^t p_{xxt}\,\mathrm{dt}\right]^{-1}
\end{aligned}
$$

and again differentiate it with respect to t knowing that $k_{xt} = 0$. As a result, we obtain the non-linear integro-differential equation for the function $p(x,t)$:

$$
\begin{aligned}
&\left[p_{xt}\int_0^t\left(\int_0^t p_{tt}\,\mathrm{dt}\right)\mathrm{dt} - p_{tt}\int_0^t\left(\int_0^t p_{xt}\,\mathrm{dt}\right)\mathrm{dt}\right]\\
&\quad\times\left[\int_0^t p_{xt}\,\mathrm{dt}\int_0^t\left(\int_0^t p_{xxt}\,\mathrm{dt}\right)\mathrm{dt} - \int_0^t\left(\int_0^t p_{xt}\,\mathrm{dt}\right)\mathrm{dt}\int_0^t p_{xxt}\,\mathrm{dt})\right]\\
&\quad-\left[\int_0^t p_{xt}\,\mathrm{dt}\int_0^t\left(\int_0^t p_{tt}\,\mathrm{dt}\right)\mathrm{dt} - \int_0^t\left(\int_0^t p_{xt}\,\mathrm{dt}\right)\mathrm{dt}\int_0^t p_{tt}\,\mathrm{dt}\right]\\
&\quad\times\left[p_{xt}\int_0^t\left(\int_0^t p_{xxt}\,\mathrm{dt}\right)\mathrm{dt} - p_{xxt}\int_0^t\left(\int_0^t p_{xt}\,\mathrm{dt}\right)\mathrm{dt}\right] = 0
\end{aligned}
$$

Transform this equation to obtain

$$
\begin{aligned}
&p_{xt}\left[\int_0^t p_{tt}\,\mathrm{dt}\int_0^t\left(\int_0^t p_{xxt}\,\mathrm{dt}\right)\mathrm{dt} - \int_0^t\left(\int_0^t p_{tt}\,\mathrm{dt}\right)\mathrm{dt}\int_0^t p_{xxt}\,\mathrm{dt}\right]\\
&+p_{tt}\left[\int_0^t\left(\int_0^t p_{xt}\,\mathrm{dt}\right)\mathrm{dt}\int_0^t p_{xxt}\,\mathrm{dt} - \int_0^t p_{xt}\,\mathrm{dt}\int_0^t\left(\int_0^t p_{xxt}\,\mathrm{dt}\right)\mathrm{dt}\right]\\
&+p_{xxt}\left[\int_0^t p_{xt}\,\mathrm{dt}\int_0^t\left(\int_0^t p_{tt}\,\mathrm{dt}\right)\mathrm{dt} - \int_0^t\left(\int_0^t p_{xt}\,\mathrm{dt}\right)\mathrm{dt}\int_0^t p_{tt}\,\mathrm{dt}\right] = 0
\end{aligned}
\tag{6.12}
$$

In equation (6.12), introduce the new unknown function $v(x,t) = p_t(x,t)/p(x,t)$. We write the following relations using the previous designations:

$$
\frac{p_{xt}}{p} = J_0 v + v_x, \qquad \frac{p_{xxt}}{p} = J_2 v + 2J_0 v_x + v_{xx}, \qquad \frac{p_{tt}}{p} = v_t + v^2.
$$

We also introduce new designations:

$$
J_{31} = \frac{1}{\varphi}\int_0^t p_{xxt}\,\mathrm{dt} = \int_0^t (1-E)(J_2 v + 2J_0 v_x + v_{xx})\,\mathrm{dt},
$$

$$J_{32} = \frac{1}{\varphi}\int_0^t \Big(\int_0^t p_{xxt}\,\mathrm{d}t\Big)\,\mathrm{d}t = \int_0^t \Big(\int_0^t (1-E)(J_2 v + 2J_0 v_x + v_{xx})\,\mathrm{d}t\Big)\,\mathrm{d}t,$$

$$J_{41} = \frac{1}{\varphi}\int_0^t p_{xt}\,\mathrm{d}t = \int_0^t (1-E)(J_0 v + v_x)\,\mathrm{d}t,$$

$$J_{42} = \frac{1}{\varphi}\int_0^t \Big(\int_0^t p_{xt}\,\mathrm{d}t\Big)\,\mathrm{d}t = \int_0^t \Big(\int_0^t (1-E)(J_0 v + v_x)\,\mathrm{d}t\Big)\,\mathrm{d}t,$$

$$J_{51} = \frac{1}{\varphi}\int_0^t p_{tt}\,\mathrm{d}t = \int_0^t (1-E)(v_t + v^2)\,\mathrm{d}t,$$

$$J_{52} = \frac{1}{\varphi}\int_0^t \Big(\int_0^t p_{tt}\,\mathrm{d}t\Big)\,\mathrm{d}t = \int_0^t \Big(\int_0^t (1-E)(v_t + v^2)\,\mathrm{d}t\Big)\,\mathrm{d}t.$$

Divide equation (6.12) by $p\varphi^2$. Then, with the new sought function $v(x,t)$, its form formally coincide with the form of equation (6.9):

$$Av_{xx} + Bv_x = Cv_t + Mv + Nv^2. \tag{6.13}$$

Here we use the following notation:

$$A = J_{41}J_{52} - J_{42}J_{51}, \qquad B = 2J_0(J_{41}J_{52} - J_{42}J_{51}) + (J_{51}J_{32} - J_{52}J_{31}),$$

$$C = J_{41}J_{32} - J_{42}J_{31}, \qquad M = J_0(J_{31}J_{52} - J_{32}J_{51}) - J_2(J_{41}J_{52} - J_{42}J_{51}),$$

$$N = C = J_{41}J_{32} - J_{42}J_{31}$$

Now consider the filtration process that follows more general non-linear law

$$\vec{v} = -\frac{k}{\mu}\,\Phi(\beta(x,y), |\nabla p|)\nabla p.$$

In the one-dimensional case this law has the form

$$\vec{v} = -\frac{k}{\mu}\,\Phi\Big(\beta(x,y), \Big|\frac{\partial p}{\partial x}\Big|\Big)\frac{\partial p}{\partial x} = kF\Big(\beta(x,y), \frac{\partial p}{\partial x}\Big),$$

where

$$F\Big(\beta(x,y), \frac{\partial p}{\partial x}\Big) = -\frac{1}{\mu}\,\Phi\Big(\beta(x,y), \Big|\frac{\partial p}{\partial x}\Big|\Big)\frac{\partial p}{\partial x}.$$

Write the filtration equation

$$\frac{\partial}{\partial x}\Big[k\,F\Big(\beta(x,y), \frac{\partial p}{\partial x}\Big)\Big] = \frac{\partial p}{\partial t}. \tag{6.14}$$

Introducing the designation $F_a \equiv \partial F/\partial a = \partial F/\partial p_x$, we rewrite the equation (6.14) as follows:

$$k(F_\beta\,\beta_x + F_a\,p_{xx}) + k_x\,F = p_t.$$

Integrate the obtained equation with respect to t from 0 to the variable upper limit t:

$$k\int_0^t (F_\beta\,\beta_x + F_a\,p_{xx})\,\mathrm{d}t + k_x\int_0^t F\,\mathrm{d}t = p - \varphi. \tag{6.15}$$

As before, we write equation (6.15) with k_x on the left-hand side:

$$k_x = \left[(p-\varphi) - k\int_0^t (F_\beta\,\beta_x + F_a\,p_{xx})\,\mathrm{d}t\right] \Big/ \int_0^t F\,\mathrm{d}t$$

and differentiate it with respect to time observing that $k_{xt} = 0$. After the transformations, we get

$$\begin{aligned} &[p_t - k(F_\beta\,\beta_x + F_a\,p_{xx})]\int_0^t F\,\mathrm{d}t \\ &\quad - \left[(p-\varphi) - k\int_0^t (F_\beta\,\beta_x + F_a\,p_{xx})\,\mathrm{d}t\right] F = 0. \end{aligned} \tag{6.16}$$

Solve equation (6.16) for the coefficient k:

$$\begin{aligned} k = &\left[F(p-\varphi) - p_t\int_0^t F\,\mathrm{d}t\right] \\ &\times\left[F\int_0^t (F_\beta\beta_x + F_a p_{xx})\,\mathrm{d}t - (F_\beta\beta_x + F_a p_{xx})\int_0^t F\,\mathrm{d}t\right]^{-1}. \end{aligned}$$

Differentiate this equality with respect to t. The condition $k_t = 0$ leads to the equation

$$\begin{aligned} &\left[F_a p_{xt}(p-\varphi) - p_{tt}\int_0^t F\,\mathrm{d}t\right] \\ &\times\left[F\int_0^t (F_\beta\beta_x + F_a p_{xx})\,\mathrm{d}t - (F_\beta\beta_x + F_a p_{xx})\int_0^t F\,\mathrm{d}t\right] \\ &-\left[F(p-\varphi) - p_t\int_0^t F\,\mathrm{d}t\right] \\ &\times\left[F_a p_{xt}\int_0^t (F_\beta\beta_x + F_a p_{xx})\,\mathrm{d}t - \right. \\ &\qquad\left. - (F_{\beta a} p_{xt}\beta_x + F_{aa} p_{xx} p_{xt} + F_a p_{xxt})\int_0^t F\,\mathrm{d}t\right] = 0. \end{aligned} \tag{6.17}$$

Thus, we again arrived at a nonlinear integro-differential equation for the function $p(x,t)$ without unknown coefficients. We transform it introducing the new unknown function $v(x,t) = p_t(x,t)/p(x,t)$ and using the notation

$$F\Big(\beta(x), \frac{\partial p}{\partial x}\Big) = F(\beta(x), \varphi J_0(1-E)), \qquad a = p_x = \varphi J_0(1-E),$$

$$\frac{p_{xt}}{p} = J_0 v + v_x, \qquad \frac{p_{tt}}{p} = v_t + v^2, \qquad p - \varphi = -\varphi E,$$

$$\frac{p_{xx}}{p} = J_2, \qquad \frac{p_{xxt}}{p} = J_2 v + 2J_0 v_x + v_{xx},$$

$$J_5 = \int_0^t F(\beta(x), \varphi J_0(1-E))\,\mathrm{d}t, \qquad J_6 = \int_0^t [F_\beta \beta_x + F_a \varphi(1-E)J_2]\,\mathrm{d}t.$$

We obtain a new non-linear integro-differential equation of the form (6.13) for the new unknown function $v(x,t)$:

$$Av_{xx} + Bv_x = Cv_t + Dv + Cv^2.$$

The coefficients of this equation are

$$A = -NJ_5F_a, \quad B = K - 2J_0J_5F_aN, \quad C = MJ_5, \quad D = -KJ_0 + J_2J_5F_aN,$$

where

$$N = \varphi[FE + J_5(1-E)v], \qquad M = FJ_6 - J_5[F_\beta\beta_x + J_2(1-E)\varphi F_a],$$

$$K = -M\varphi EF_a + NJ_6F_a - NJ_5F_{\beta a}\beta_x - NJ_2J_5(1-E)\varphi F_{aa}.$$

6.3. THE QUASI-INVERSION PROBLEM FOR THE CONSIDERED CASES

So, after eliminating the unknown coefficient from the corresponding equation, we have to study the equation

$$Av_{xx} + Bv_x = Cv_t + Mv + Nv^2,$$

whose coefficients depend on the integrals containing the unknown function. To formulate the corresponding quasi-inversion problem, it is necessary to construct an operator conjugate to the non-linear integro-differential operator

$$Pv = -Av_{xx} - Bv_x + Cv_t + Dv,$$

where $D = M + Nv$. Then, using the described technique, we turn to constructing an operator conjugate to the linear differential operator

$$Pv^{(n)} = -A^{(n-1)}v_{xx}^{(n)} - B^{(n-1)}v_x^{(n)} + C^{(n-1)}v_t^{(n)} + D^{(n-1)}v^{(n)}. \tag{6.18}$$

To this end, we use the usual rules (see Sobolev, 1966; Friedman, 1968). As a result, the operator conjugate to the operator (6.18) takes the form

$$P^*v = -(Av)_{xx} + (Bv)_x - (Cv)_t + Dv.$$

Now the equation of the quasi-inversion method

$$\frac{1}{\varepsilon_1} P^* P\tilde{v} + P\tilde{v} - \varepsilon_2 \tilde{v}_{tt}$$

can be written in the following expanded form:

$$\begin{aligned}
&\frac{1}{\varepsilon_1} A^2 \tilde{v}_{xxxx} + \frac{4}{\varepsilon_1} AA_x \tilde{v}_{xxx} \\
&+ \Big\{\frac{1}{\varepsilon_1}[2A_x^2 + 2AA_{xx} - B^2 + (AB)_x + (AC)_t - 2AD] - A\Big\}\tilde{v}_{xx} \\
&+ \Big\{\frac{1}{\varepsilon_1}[-2BB_x + (BC)_t + (AB)_{xx} - 2(AD)_x] - B\Big\}\tilde{v}_x \\
&+ \Big\{\frac{1}{\varepsilon_1}[-2CC_t - (AC)_{xx} + (BC)_x] + C\Big\}\tilde{v}_t \\
&- \Big(\frac{1}{\varepsilon_1} C^2 + \varepsilon_2\Big)\tilde{v}_{tt} \\
&+ \frac{1}{\varepsilon_1}[-2(AC)_x + 2BC]\tilde{v}_{tx} \\
&+ \Big\{\frac{1}{\varepsilon_1}[-(AD)_{xx} + (BD)_x - (CD)_t + D^2] + D\Big\}\tilde{v} = 0.
\end{aligned} \tag{6.19}$$

Equation (3.11) is a special case of equation (6.19) with $D = 0$. The problem formulation is similar to that in Chapter 3 and the study of it is carried out following the same procedure: equation (6.19) is simplified, then the quasi-inversion problem corresponding to the original problem is formulated and solved numerically by a finite-difference method. The coefficients A, B, C, D in equation (6.19) are chosen in accordance with the corresponding case under consideration.

Summary

In Lattes and Lions (1970), the authors note in conclusion:

"In the class of "inverse problems" there are problems in which the coefficients of differential operators are unknown, whereas the general forms of these operators are known. These unknown coefficients can be either 1) the operator coefficients "inside the domain" or 2) the coefficients of boundary operators.

In case 1) several method can be proposed: the differential approximation method ... accompanied by various iteration methods ..., the method of "linear forms" ..., and reduction to an integral equation.

Here we only wish to note that the quasi-inversion method is applicable for solving inverse problems in case 2)".

The research carried out by the author showed that the quasi-inversion method can be applied also for solution of "interior coefficient inverse problems" by reducing them to the problem of continuation of a solution to a parabolic equation. This reduction is based on the results of the proofs of the uniqueness theorems for solutions of the corresponding coefficient inverse problems (see Klibanov, 1984a,b, 1986; Klibanov and Danilaev, 1990).

When studying the potential of the quasi-inversion method, the author has not confined himself to considering only model examples. In Danilaev (1993) and Danilaev, Gortyshov, and Kuz'min (1988) this method was used in studying a practical problem of continuation of a solution to a parabolic equation (in designing a power installation component). The calculations also produced good results in comparison with the results of direct high-precision measurements.

Acknowledgement

I am grateful to all who assisted me in working on this book for helpful advice aimed at its improvement. Professor V. Ya. Bulygin was the first who advised me to study the problem of determining filtration parameters using the methods of mathematical modelling. Professor G. G. Tumashev pointed out some interesting possibilities for application of the quasi-inversion method. I also acquired a very valuable experience working in contact with Academician M. M. Lavrent'ev, V. G. Romanov, corresponding member of the Russian Academy of Sciences, and their pupils while being an invited researcher at the Institute of Mathematics, Siberian Branch of the Russian Academy of Sciences, and participating in the seminars and international conferences on inverse problems held by Moscow Institute of Aircraft Engineering and Moscow State University under the direction of Professor O. M. Alifanov.

I wish to thank the scientific editor, Professor A. M. Elizarov, for his numerous critical remarks, and my referees, Professor P. N. Vabishchevich and Professor V. A. Morozov. Working in contact with them helped me to get a broader view of the studied problem and to improve the book substantially.

I would also like to thank Soros grant winners, President of the Kazan Mathematical Society, Professor A. V. Lapin and Chairman of the Department of Special Mathematics at A. N. Tupolev Kazan State Technical University, Honoured Researcher in Science and Technology of Tatarstan, Professor K. G. Garaev, who supported the publication of this book.

Bibliography

Abasov, M. T., Azimov, E. H., and Ibragimov, T. M. (1991). On solution of the coefficient inverse problem for the non-stationary oil and gas filtration in seam. *Dokl. Akad. Nauk SSSR* 318 (3), 566–569 (in Russian).

Alifanov, O. M. (1979). *Identification of Heat Exchange Processes of Flying Vehicles.* Mashinostroenie, Moscow (in Russian).

Alifanov, O. M. (1988). *Inverse Problems of Heat Exchange.* Mashinostroenie, Moscow (in Russian).

Alifanov, O. M., Artyukhin, E. A., and Rumyantsev, S. V. (1988). *Extremal Methods for Solving Ill-Posed Problems.* Nauka, Moscow (in Russian).

Alifanov, O. M. and Klibanov, M. V. (1985). Uniqueness conditions and the method of solution of the coefficient inverse problem of heat conduction. *Inzh.-Fiz. Zh.* 48 (6), 998–1003 (in Russian).

Bakushinskii, A. B. and Goncharskii, A. V. (1989). *Ill-Posed Problems. Numerical Methods and Applications.* Moscow State University, Moscow (in Russian).

Banks, H. T. and Lamm, P. D. (1985). Estimation of variable coefficients in parabolic distributed systems. *IEEE Trans. Automat. Control* 30, 386–398.

Beck, J., Blackwell, B., and Saint Clair, Ch., jr. (1989). *Ill-Posed Inverse Problems of Heat Conduction.* Mir, Moscow (in Russian).

Bellman, R., Gluss, B., and Roth, R. (1965). Segmental differential approximation and the "black box" problem. *J. Math. Anal. and Appl.* 12, 91–104.

Bellman, R., Kagiwada, H., Kalaba, R., and Ueno, S. (1965). *Inverse Problems in Radioactive Transfer: Layered Media.* Icarus 4.

Belov, V. and Lavrent'ev, M., jr. (1996). One inverse problem for the heat equation. *J. Inv. Ill-Posed Problems* 4 (6), 499–511.

Bolgarskii, A. V., Muhachev, G. A., and Shchyukin, V. K. (1975). *Thermodynamics and Heat Transfer.* Vysshaya Shkola, Moscow (in Russian).

Bukhgeim, A. L. (1986). *Difference Methods for Solving Ill-Posed Problems.* Computing Center, Siberian Branch of the USSR Acad. Sci., Novosibirsk (in Russian).

Bukhgeim, A. L. (2000). *Introduction to the Theory of Inverse Problems.* VSP, Utrecht.

Bukhgeim, A. L. and Klibanov, M. V. (1981). Uniqueness in the large of a class of many-dimensional inverse problems. *Dokl. Akad. Nauk SSSR* 260 (2), 269–272 (in Russian).

Bulygin, V. Ya. (1958). Hydromechanical analysis of an oil seam functioning based on the data of technical specifications. In: *The Works on Filtration Theory.* Uch. Zap. Kazan. Univ. 118(2). Kazan, 4–67 (in Russian).

Bulygin, V. Ya. (1974). *Hydromechanics of an Oil Seam.* Nedra, Moscow (in Russian).

Bulygin, V. Ya. and Danilaev, P. G. (1971). On determining water permeability by solving the improperly stipulated system of linear algebraic equations. *Numerical methods in technical and economic calculations.* Kazan State University, Kazan, 15–18 (in Russian).

Butkovskii, A. G. (1975). *Control Methods for Systems with Distributed Parameters.* Nauka, Moscow (in Russian).

Buzinov, S. N. and Umrikhin, I. D. (1984). *Study of Oil and Gas Wells and Seams.* Nedra, Moscow (in Russian).

Cahen, G. (1963). Détermination expérimentale des paramétres des systéms à retard. *Revue Francaise Fraitement Information* (1), 15–23.

Charnyi, M. N. (1997). *Underground Fluid Dynamics.* Gostoptekhizdat, Moscow (in Russian).

Chavent, G. (1970). Une méthode de résolution de problème inverse dans les équations aux dérivées partiélles. *Bulletin de l'Academie Polonaise des Sciences, Série des Sciences Techniques* 18 (8), 99–105.

Chavent, G. (1971). Analyse fonctionelle et identification de coefficients répartis dans les équations aux dérivées partiélles. *These d'État, Faculté des Sciences de Paris.*

Chavent, G., Dupuy, M., and Lemonnier, P. (1975). History matching by use of optimal control theory. *Soc. Pet. Eng.* 15, 74–86.

Danilaev, P. G. (1978). On calculation of water permeability of an exploited oil seam. *University Bulletin. Oil and Gas* (2), 51–54 (in Russian).

Danilaev, P. G. (1980). On the statement of some ill-posed problems of gas dynamics. In: *Gas Dynamics in Engines of Flying Vehicles.* Kazan Institute of Aviation, Kazan, 3–9 (in Russian).

Danilaev, P. G. (1981). On some inverse problems of heat exchange in solid-propellant jet engines. In: *Gas Dynamics in Engines of Flying Vehicles.* Kazan Institute of Aviation, Kazan, 9–14 (in Russian).

Danilaev, P. G. (1986). Experience of numerical solution of the one-dimensional evolutionary inverse problem by the quasi-inversion method. *Numerical Methods of Continuum Mechanics* 17 (5), 69–76 (in Russian).

Danilaev, P. G. (1987a). Experience of numerical solution of the one-dimensional evolutionary inverse problem by the quasi-inversion method using implicit finite-difference schemes. *Modelirovanie v Mekhanike* 1 (18), 42–48 (in Russian).

Danilaev, P. G. (1987b). Self-similar solution to the problem of determining a water permeability of an inhomogeneous oil seam. In: *Optimization of Oil Production and Problems of Underground Hydromechanics.* Kazan Filiation of the USSR Acad. Sci., Kazan, 39–43 (in Russian).

Danilaev, P. G. (1988). Solution of inverse problems of heat conduction by the quasi-inversion method. In: *Proceedings of VI All-Union Seminar "Inverse Problems and Identification of Heat Exchange Processes"*. Moscow, 32–33 (in Russian).

Danilaev, P. G. (1989a). Numerical solution of a one-dimensional parabolic equation with non-standard initial and boundary conditions. *Modelirovanie v Mekhanike* 3 (20), 61–68 (in Russian).

Danilaev, P. G. (1989b). Numerical solution of the problem of determining the seam pressure in non-classical statement. In: *Mathematical Modelling of the Processes of Filtration and Optimization in Oil Production.* Kazan Filiation of the USSR Acad. Sci., Kazan, 29–33 (in Russian).

Danilaev, P. G. (1993). Numerical solution of the internal inverse problem for parabolic equations and its application. *Modelirovanie v Mekhanike* 7 (24), 45–50 (in Russian).

Danilaev, P. G. (1996). On solution of coefficient inverse problems by the quasi-inversion method and their applications. *II Siberian Congress on Applied and Industrial Mathematics: Abstracts.* Novosibirsk (in Russian).

Danilaev, P. G. (1997). Numerical solution of the coefficient inverse problem for a parabolic equation of a special form. *Bulletin of A. N. Tupolev Kazan State Technical University* 2, 26–28 (in Russian).

Danilaev, P. G. (1998a). Solution of coefficient inverse problems of underground fluid dynamics by the quasi-inversion method. *III International Conference "Identification of Dynamic Systems and Inverse Problems"*. Moscow - St. Petersburg, 211–218 (in Russian).

Danilaev, P. (1998b). On the filtration non-homogeneous porous stratum parameters identification problem. *The International Symposium on Inverse Problems in Engineering Mechanics (ISIP'98).* Nagano City.

Danilaev, P. G. and Golubev, G. V. (1994a). Solution of coefficient inverse problems of heat conduction and their applications. In: *Heat Conduction and Thermal Insulation: Proceedings of I Russian National Conference on Heat Exchange. V. 10, Pt 1.* Moscow Institute of Power Engineering, Moscow, 69–74 (in Russian).

Danilaev, P. G. and Golubev, G. V. (1994b). On solution of coefficient inverse problems and their applications. In: *Identification of Dynamic Systems and Inverse Problems: Proceedings of II International Conference. B7.* St. Petersburg, 1–12 (in Russian).

Danilaev, P. G., Gortyshov, Yu. F., and Kuz'min, A. P. (1988). Numerical solution of the non-classical problem of calculation of one-dimensional nonstationary temperature field. In: *Heat Exchange and Friction in Engines and Power Plants of Flying Vehicles.* Kazan Institute of Aviation, Kazan, 12–16 (in Russian).

Development of Research in Filtration Theory in the USSR (1917–1967). (1969). Nauka, Novosibirsk (in Russian).

Dmitriev, V. I. (Ed) (1982). *Computational Mathematics and Computing Systems in Exploration Geophysics. Reference Book on Geophysics.* Nedra, Moscow (in Russian).

Douglas, J., jr. and Jones, S. (1962). The determination of a coefficient in a parabolic differential equation. *J. Math. and Mech.* 2, 919–926.

Evdokimov, Yu. K. (1995). *Distributed measuring media.* Synopsis of Ph.D. Thesis. Kazan State Technical University, Kazan (in Russian).

Friedman, A. (1968). *Partial Differential Equations of Parabolic Type.* Mir, Moscow (in Russian).

Golubev, G. V. (1992). *Analytic and Numerical Methods for Determination of Filtration Parameters and Pressure for Fluid and Gas Motion in the Inhomogeneous Porous Medium.* Synopsis of Ph.D. Thesis. Kazan State University, Kazan (in Russian).

Golubev, G. V. and Danilaev, P. G. (1981). On the problem of determination of filtration parameters. *Dynamics of Multi-Phase Media: Proceedings of V All-Union Seminar "Numerical Methods for Solving the Problems of Filtration of Multi-Phase Incompressible Fluid".* Novosibirsk, 133–139 (in Russian).

Golubev, G. V. and Danilaev, P. G. (1983). On the problem of determination of filtration parameters of the seam for non-linear filtration. *Dynamics of Multi-Phase Media: Proceedings of VI All-Union Seminar "Numerical Methods for Solving the Problems of Filtration of Multi-Phase Incompressible Fluid"*. Novosibirsk, 116–121 (in Russian).

Golubev, G. V. and Danilaev, P. G. (1987). Numerical methods of determination of filtration parameters and solution of inverse problems. In: *Numerical Methods for Solving the Problems of Filtration of Multi-Phase Incompressible Fluid: Collection of Scientific Works of the Institute of Theoretical and Applied Mechanics, Siberian Branch of the USSR Acad. Sci.* Novosibirsk, 104–108 (in Russian).

Golubev, G. V. and Danilaev, P. G. (1990). On determining the fields of filtration parameters of exploited seams by the results of measurements. In: *Development of Gas-Condensate Fields: Proceedings of International Conference, Workshop 3*. Krasnodar, 169–174 (in Russian).

Golubev, G. V. and Danilaev, P. G. (1991a). On some inverse problems of filtration with initial gradient of pressure. *Filtration of Multi-Phase Systems: Proceedings of the All-Union Seminar*. Novosibirsk, 71–76 (in Russian).

Golubev, G. V. and Danilaev, P. G. (1991b). Determination of filtration parameters of one- and two-phase fluid motion in the porous medium. *Mechanics of Fluid and Multi-Phase Media*. Filiation of the USSR Acad. Sci., Tashkent, 17–23 (in Russian).

Golubev, G. V. and Danilaev, P. G. (1992a). Application of numerical methods of solution of quasi-linear equations to the problems of fluid and gas motion in the inhomogeneous porous medium. *Modelirovanie v Mekhanike* 6 (4), 13–20 (in Russian).

Golubev, G. V. and Danilaev, P. G. (1992b). Inverse problems in non-linear filtration. *Flow Through Porous Media: Fundamentals and Reservoir Engineering Applications*. Inst. for Problems in Mech., Russian Acad. Sci., Moscow.

Golubev, G. V. and Danilaev, P. G. (1996a). On the problem of determining the filtration-capacitive parameters of seams of exploited oil-fields. In: *Fundamental Problems of Oil and Gas: All-Russian Scientific Conference. Reports and Speeches* 4, 281–291 (in Russian).

Golubev, G. V. and Danilaev, P. G. (1996b). Determination of filtration-capacitive parameters of exploited seams in high-viscous oil production. *Oil and Bitumen: Proceedings of International Conference "Problems of Complex Development of Hardly Extractable Reserves of Oil and Natural Bitumen"* 6, 1921–1929 (in Russian).

Golubev, G. V., Danilaev, P. G., and Tumashev, G. G. (1978). *Determination of Water Permeability of Inhomogeneous Oil Seams by Non-Local Methods.* Kazan State University, Kazan (in Russian).

Golubev, G. V. and Tumashev, G. G. (1972). *Filtration of Incompressible Fluid in the Inhomogeneous Porous Medium.* Kazan State University, Kazan (in Russian).

Iskenderov, A. D. (1971). Boundary inverse problems for determining the parameters of filterable media. *Izv. Akad. Nauk AzSSR. Ser. Fiz.-Mat. i Tekhn. Nauk* (2), 30–34 (in Russian).

Ivanov, V. K., Vasin, V. V., and Tanana, V. P. (1978). *The Theory of Linear Ill-Posed Problems and its Applications.* Nauka, Moscow (in Russian).

Kagivada, H. (1964). *These U.C.L.A.*

Khairullin, M. H. (1986). On solution of inverse problems of underground hydromechanics by the regularizing algorithms in the sense of A. N. Tikhonov. *Zh. Vych. Mat. i Mat. Fiz.* 26 (5), 780–783 (in Russian).

Khairullin, M. H. (1988). On regularization of the coefficient inverse problem of non-stationary filtration. *Dokl. Akad. Nauk* 299 (5), 1108–1111 (in Russian).

Klibanov, M. V. (1984a). Inverse problems in the large and Carleman's estimates. *Differents. Uravn.* 20 (6), 1035–1041 (in Russian).

Klibanov, M. V. (1984b). Uniqueness in the large in inverse problems for one class of differential equations. *Differents. Uravn.* 20 (11), 1947–1953 (in Russian).

Klibanov, M. V. (1986). Uniqueness of solution of two inverse problems for the Maxwell system. *Zh. Vych. Mat. i Mat. Fiz* 26 (7), 1063–1071 (in Russian).

Klibanov, M. V. and Danilaev, P. G. (1990). On solution of coefficient inverse problems by the quasi-inversion method. *Dokl. Akad. Nauk SSSR* 310 (3), 528–532 (in Russian).

Kozdoba, L. A. (1975). *Methods for Solving the Non-Linear Problems of Heat Conduction.* Nauka, Moscow (in Russian).

Kozdoba, L. A. (1992). *Computational Thermophysics.* Naukova Dumka, Kiev (in Russian).

Kozdoba, L. A. and Krukovskii, P. G. (1982). *Methods for Solving Inverse Problems of Heat Transfer.* Naukova Dumka, Kiev (in Russian).

Kravaris, C. and Sienfeld, J. H. (1982). Distributed parameter identification in geophysics-petroleum reservoirs and aquifers. In: *Distributed Parameter Control Systems.* S. Tzafestas, Ed. Pergamon, New York.

Kravaris, C. and Sienfeld, J. H. (1985). Identification of parameters in distributed parameter systems by regularization. *SIAM J. Control and Optimization* 23 (2), 217–241.

Kravaris, C. and Sienfeld, J. H. (1986). Identification of spatially varying parameters in distributed parameters systems by discrete regularization. *J. of Math. Analysis and Applications* 119, 1128–152.

Krikunov, Yu. M. (1970). *Lectures in the Equations of Mathematical Physics and Integral Equations.* Kazan State University, Kazan (in Russian).

Kurpisz, K. and Novak, A. J. (1995). *Inverse Thermal Problems.* Southampton, UK and Boston, USA: Computational Mechanics Publication.

Ladyzenskaja, O. A., Solonnikov, V. A., and Ural'ceva, N. N. (1967). *Linear and Quasi-Linear Equations of Parabolic Type.* Nauka, Moscow (in Russian).

Lattès, R. and Lions, J.-L. (1967). *Méthode de Quasi-réversibilité et Applications.* Dunod, Paris.

Lavrent'ev, M. M. (1981). *Ill-Posed Problems for Differential Equations.* Novosibirsk State University, Novosibirsk (in Russian).

Lavrent'ev, M. M., Reznitskaya, K. G., and Yakhno, V. G. (1982). *One-Dimensional Inverse Problems of Mathematical Physics.* Nauka, Novosibirsk (in Russian).

Lavrent'ev, M. M., Romanov, V. G., and Shishatskii, S. P. (1980). *Ill-Posed Problems of Mathematical Physics and Analysis.* Nauka, Moscow (in Russian).

Lavrent'ev, M. M., Vasil'ev, V. G., and Romanov, V. G. (1969). *Many-Dimensional Inverse Problems for Differential Equations.* Nauka, Novosibirsk (in Russian).

Liskovets, O. A. (1981). *Variational Methods for Solution of Unstable Problems.* Nauka i Tekhnika, Minsk (in Russian).

Makarov, A. M. and Romanovskii, M. R. (1978). Solution of coefficient inverse problems by the method of regularization using a spline-function. *Inzh.-Fiz. Zh.* 34 (2), 332–337 (in Russian).

Morozov, V. A. (1987a). *Methods of Regularization of Unstable Problems.* Moscow State University, Moscow (in Russian).

Morozov, V. A. (1987b). *Regular Algorithms of Solution of Ill-Posed Problems.* Nauka, Moscow (in Russian).

Muzylev, N. V. (1977). On the quasi-inversion method. *Zh. Vych. Mat. i Mat. Fiz.* 17 (3), 556–561 (in Russian).

Popov, Yu. P. and Samarskii, A. A. (1988). Computational experiment. In: *Computers, Models, and Experiment.* Nauka, Moscow (in Russian).

Romanov, V. G. (1969). *Inverse Problems for Parabolic Equations.* Nauka, Novosibirsk (in Russian).

Romanov, V. G. (1984). *Inverse Problems of Mathematical Physics.* Nauka, Moscow (in Russian).

Romanov, V. G., Kabanikhin, S. I., and Pukhnacheva, T. P. (1984). *Inverse Problems of Electrodynamics.* Computing Center of Siberian Branch of the USSR Acad. Sci., Novosibirsk (in Russian).

Samarskii, A. A. (1977). *The Theory of the Difference Schemes.* Nauka, Moscow (in Russian).

Samarskii, A. A. (1979). Mathematical Modelling and Computational Experiment. *Vestnik Akad. Nauk SSSR* (5), 38–49 (in Russian).

Samarskii, A. A. and Nikolaev, E. S. (1978). *Methods of Solving Difference Equations.* Nauka, Moscow (in Russian).

Samarskii, A. A. and Vabishchevich, P. N. (1990). Difference schemes for unstable problems. *Mat. Modelirovanie* 2 (11), 89–98 (in Russian).

Samarskii, A. A. and Vabishchevich, P. N. (1997). Difference methods for solution of inverse problems of mathematical physics. In: *Fundamentals of Mathematical Modelling.* Nauka, Moscow (in Russian).

Savateev, E. G. (1995). On problems of determining the source function in a parabolic equation. *J. Inv. Ill-Posed Problems* 3 (1), 83–102.

Sedov, L. I. (1967). *Methods of Similarity and Dimension in Mechanics.* Nauka, Moscow (in Russian).

Shamsiev, M. N. (1997). *Numerical Methods of Solution of Inverse Problems for Saturated Porous Media.* Synopsis of Ph.D. Thesis. Ufa (in Russian).

Sobolev, S. L. (1966). *Equations of Mathematical Physics.* Nauka, Moscow (in Russian).

Tamme, E. E. (1972). On the stability of difference schemes for solution of ill-posed problems by the quasi-inversion method. *Zh. Vych. Mat. i Mat. Fiz.* 12 (5), 1319–1325 (in Russian).

Tikhonov, A. N. (1965a). On the ill-posed problems of linear algebra and stable method for their solutions. *Dokl. Akad. Nauk* 163 (6), 591–595 (in Russian).

Tikhonov, A. N. (1965b). On the stability of algorithms for solving degenerate systems of linear algebraic equations. *Zh. Vych. Mat. i Mat. Fiz.* 5 (4), 718–722 (in Russian).

Tikhonov, A. N. and Arsenin, V. Ya. (1979). *Methods of Solving Ill-Posed Problems.* Nauka, Moscow (in Russian).

Tikhonov, A. N., Goncharskii, A. V., Stepanov, V. V., and Yagola, A. G. (1983). *Regularizing Algorithms and A Priori Information.* Nauka, Moscow (in Russian).

Tikhonov, A. N. and Samarskii, A. A. (1961). On homogeneous difference schemes. *Zh. Vych. Mat. i Mat. Fiz.* 1 (1), 5–63 (in Russian).

Tikhonov, A. N. and Samarskii, A. A. (1966). *Equations of Mathematical Physics.* Nauka, Moscow (in Russian).

Tsirel'man, N. M. (1984). On the well-posedness of the analytical solution to the inverse problem of heat conduction. In: *Inverse Problems and Identification of Heat Exchange: Abstracts. V All-Union Seminar.* Ufa Institute of Aviation, Ufa (in Russian).

Vabishchevich, P. N. (1991a). *Quasi-Inversion Method for Approximate Solution of the Inverse Problems of Heat Exchange.* Preprint N 11. Institute of the Problems of Safe Development of Atomic Power Engineering, the USSR Acad. Sci., Moscow (in Russian).

Vabishchevich, P. N. (1991b). *Quasi-Inversion Method for Evolutionary Equations of the Second Kind.* Preprint N 26. Computing Center of Mathematical Modelling, the USSR Acad. Sci., Moscow (in Russian).

Vabishchevich, P. N. (1991c). *Difference Schemes of the Quasi-Inversion Method for Evolutionary Equations of the Second Kind.* Preprint N 25. Computing Center of Mathematical Modelling, the USSR Acad. Sci., Moscow (in Russian).

Vabishchevich, P. N. and Denisenko, A. Yu. (1990). Numerical method for solving coefficient inverse problems. *Methods for Mathematical Modelling and Computational Diagnostics.* Moscow State University, Moscow (in Russian).

Bei Fragen zur Produktsicherheit wenden Sie sich bitte an:
If you have any questions regarding product safety,
please contact:

Walter de Gruyter GmbH
Genthiner Straße 13
10785 Berlin
productsafety@degruyterbrill.com